MOUNTAIN BILLY UNDER THE GNATCATCHER'S OAK

A-BIRDING ON A BRONCO

BY

FLORENCE A. MERRIAM

I do invite you . . . to my house . . .
after, we'll a-birding together.
SHAKESPEARE.

ILLUSTRATED

APPLEWOOD BOOKS
Carlisle, Massachusetts

A-Birding on a Bronco was originally published in 1896
by Houghton, Mifflin and Company.

Thank you for purchasing an Applewood book. Applewood reprints America's lively classics—books from the past that are still of interest to modern readers. Our mission is to tell the authentic stories of people, places, ideas, and history. In order to preserve the authenticity of the works we publish, we do not alter the content in the editions we reissue. Sometimes words, images, or ideas from the past will seem inappropriate to the modern reader. We believe in the value of presenting the past, just as it was, for the purposes of informing the present and future.

For a complete list of books currently available,
please visit us at www.awb.com

Applewood Books
An imprint of Arcadia Publishing

978-1-4290-9695-9

1 2 3 4 5 6 7 8 9 10

MANUFACTURED IN THE UNITED STATES OF AMERICA
WITH AMERICAN-MADE MATERIALS

PREFATORY NOTE.

THE notes contained in this book were taken from March to May, 1889, and from March to July, 1894, at Twin Oaks in southern California. Twin Oaks is the post-office for the scattered ranch-houses in a small valley at the foot of one of the Coast Ranges, thirty-four miles north of San Diego, and twelve miles from the Pacific.

As no collecting was done, there is doubt about the identity of a few species; and their names are left blank or questioned in the list of birds referred to in the text. In cases where the plumage of the two sexes is practically identical, and only slight mention is made of the species, the sexes have sometimes been arbitrarily distinguished in the text.

Several of the articles have appeared before, in somewhat different form, in 'The Auk,' 'The Observer,' and 'Our Animal Friends;' all the others are published here for the first time.

The illustrations are from drawings of birds and nests by Louis Agassiz Fuertes, and from photographs taken in the valley; together with some of eucalyptus-trees from Los Angeles, for the use of which I am indebted to the courtesy of Dr. B. E. Fernow, Chief of the Division of Forestry of the U. S. Department of Agriculture.

In the preparation of the book I have been kindly assisted by Miss Isabel Eaton, and have received from my brother, Dr. C. Hart Merriam, untiring criticism and advice.

FLORENCE A. MERRIAM.

LOCUST GROVE, N. Y.,
July 15, 1896.

CONTENTS.

LIST OF ILLUSTRATIONS.

BIRDS REFERRED TO IN THE TEXT.[1]

White Egret. *Ardea egretta.*
Green Heron. *Ardea virescens anthonyi.*
Spotted Sandpiper. *Actitis macularia.*
Valley Quail. *Callipepla californica vallicola.*
Mourning Dove. *Zenaidura macroura.*
Turkey Vulture. *Cathartes aura.*
Hawk. *Buteo* ——.
Sparrow Hawk. *Falco sparverius deserticolus.*
American Barn Owl. *Strix pratincola.*
Western Horned Owl. *Bubo virginianus subarcticus.*
Burrowing Owl. *Speotyta cunicularia hypogæa.*
Road-runner. *Geococcyx californianus.*
California Woodpecker. *Melanerpes formicivorus bairdi.*
Red shafted Flicker. *Colaptes cafer.*
Dusky Poor-will, *Phalænoptilus nuttalli californicus.*
Black-chinned Hummingbird. *Trochilus alexandri.*
Rufous Hummingbird. *Selasphorus rufus.*
Arkansas Kingbird. *Tyrannus verticalis.*
Cassin's Kingbird. *Tyrannus vociferans.*
Black Phœbe. *Sayornis nigrescens.*
Western Wood Pewee. *Contopus richardsonii.*
Flycatcher. *Empidonax* ——.
Horned Lark. *Otocoris alpestris chrysolæma.*
California Jay. *Aphelocoma californica.*
American Crow. *Corvus americanus.*
Yellow-headed Blackbird. *Xanthocephalus xanthocephalus.*
Red-winged Blackbird. *Agelaius phœnicius* ——.
Arizona Hooded Oriole. *Icterus cucullatus nelsoni.*
Bullock's Oriole. *Icterus bullocki.*

[1] In classification and nomenclature this list conforms to the American Ornithologists' Union 'Check-List of North American Birds,' Second Edition, 1895. L. S. Foster, New York.

Brewer's Blackbird. *Scholocophagus cyanocephalus.*
Western House Finch. *Carpodacus mexicanus frontalis.*
Goldfinch. *Spinus ——.*
White-crowned Sparrow. *Zonotrichia leucophrys gambeli* (?).
Golden-crowned Sparrow. *Zonotrichia coronata.*
Heerman's Song Sparrow. *Melospiza fasciata heermanni* (?).
Spurred Towhee or Chewink. *Pipilo maculatus megalonyx.*
Green-tailed Towhee. *Pipilo chlorurus.*
California Towhee. *Pipilo fuscus crissalis.*
Black-headed Grosbeak. *Habia melanocephala.*
Western Blue Grosbeak. *Guiraca cærulea eurhyncha.*
Lazuli Bunting. *Passerina amœna.*
Louisiana Tanager. *Piranga ludoviciana.*
Cliff Swallow. *Petrochelidon lunifrons.*
Phainopepla. *Phainopepla nitens.*
White-rumped Shrike. *Lanius ludovicianus excubitorides.*
Warbling Vireo. *Vireo gilvus* (?).
Hutton's Vireo. *Vireo huttoni* (?).
Least Vireo. *Vireo bellii pusillus* (?).
Long-tailed Chat. *Icteria virens longicauda.*
American Pipit. *Anthus pensilvanicus.*
California Thrasher. *Harporhynchus redivivus.*
Vigors's Wren. *Thryothorus bewickii spilurus.*
Western House Wren. *Troglodytes œdon aztecus.*
Plain Titmouse. *Parus inornatus.*
Wren-tit. *Chamæa fasciata.*
California Bush-tit. *Psaltriparus minimus californicus.*
Western Gnatcatcher. *Polioptila cærulea obscura.*
Varied Thrush or Oregon Robin. *Hesperocichla nævia.*
Western Bluebird. *Sialia mexicana occidentalis.*

A-BIRDING ON A BRONCO.

I.

OUR VALLEY.

"CLIMB the mountain back of the house and you can see the Pacific," the ranchman told me with a gleam in his eye; and later, when I had done that, from the top of a peak at the foot of the valley he pointed out the distant blue mountains of Mexico. Then he gave me his daughter's saddle horse to use as long as I was his guest, that I might explore the valley and study its birds to the best advantage. Before coming to California, I had known only the birds of New York and Massachusetts, and so was filled with eager enthusiasm at thought of spending the migration and nesting season in a new bird world.

I had no gun, but was armed with opera-glass and note-book, and had Ridgway's Manual to turn to in all my perplexities. Every morning, right after breakfast, my horse was brought to the door and I set out to make the rounds of the valley. I rode till dinner time, getting acquainted with the migrants as they came from the south, and calling at the more distant nests on the way. After dinner I would take my camp-stool and stroll through the oaks at the head of the valley, for a quiet study of the nearer nests. Then once more my horse would be brought up for me to take a run before sunset; and at night I would identify my new birds and write up the notes of the day. What more could observer crave? The world was mine. I never spent a happier spring. The freedom and novelty of ranch life and the exhilaration of days spent in the saddle gave added zest to the delights of a new fauna. In my small valley circuit of a mile and a half, I made the acquaintance of about seventy-five birds, and without resort to the gun was able to name fifty-six of them.

My saddle horse, a white bronco who went by the musical name of Canello, had been broken by a Mexican whose cruelty had tamed the wild blood in his veins and left him with a fear of all swarthy skins. Now he could be ridden bareback by the little girls, with only a rope noose around his nose, and was warranted to stand still before a flock of birds so long as there was grass to eat. He was to be relied on as a horse of ripe experience and mature judgment in matters of local danger. No power of bit or spur could induce him to set foot upon a piece of 'boggy land,' and to give me confidence one of the ranchman's sons said, "Wherever I've killed a rattlesnake from him he 'll shy for years;" and went on to cite localities where a sudden, violent lurch had nearly sent him over Canello's head! What greater recommendation could I wish?

If the old horse had had any wayward impulses left, his Mexican bit would have subdued them. It would be impossible to use such an iron in the mouth of an eastern horse. They say the Mexicans sometimes break horses' jaws with it. From the middle of the bit, a flat bar of iron, three quarters of an inch wide, extended back four inches, lying on the horse's tongue or sticking into the roof of his mouth, according to the use of the curb — there was no other rein. The bit alone weighed sixteen ounces. The bridle, which came from Enseñada in Lower California, then the seat of a great gold excitement, was made of braided raw-hide. It was all hand work; there was not a buckle about it. The leather quirt at the end of the reins was the only whip necessary. When I left the ranch the bridle was presented to me, and it now hangs behind my study door, a proud trophy of my western life, and one that is looked upon with mingled admiration and horror by eastern horsemen.

Canello and I soon became the best of friends. I found in him a valuable second — for, as I had anticipated, the birds were used to grazing horses, and were much less suspicious of an equestrian than a foot passenger — and he found in me a movable stake, constantly leading him to new grazing ground; for when there was a nest to watch I simply hung the bridle over the pommel and let him eat, so getting free hands for opera-glass and note-book. To be sure, there were slight causes of difference between us. He liked to watch birds in the high alfalfa under the sycamores, but when it came to standing still where the hot sun beat down through the brush and there was nothing to eat, his interest in ornithology flagged perceptibly. Then he sometimes carried the rôle of grazing horse too far, marching off to a fresh clump of grass out of sight of my nest at the most interesting moment; or when I was intently gazing through my glass at a rare bird, he would sometimes give a sudden kick at a horse-fly, bobbing the glass out of range just as I was making out the character of the wing-bars.

From the ranch-house, encircled by live-oaks, the valley widened out, and was covered with orchards and vineyards, inclosed by the low brush-grown ridges of the Coast Mountains. It was a veritable paradise for the indolent field student. With so much insect-producing verdure, birds were everywhere at all times. There were no long hours to sit waiting on a camp-stool, and only here and there a treetop to 'sky' the wandering birds. The only difficulty was to choose your intimates.

Canello and I had our regular beat, down past the blooming quince and apricot orchard, along

OUR VALLEY

the brush-covered side of the valley where the migrants flocked, around the circle through a great vineyard in the middle of the valley, past a pond where the feathered settlers gathered to bathe, and so back home to the oaks again.

I liked to start out in the freshness of the morning, when the fog was breaking up into buff clouds over the mountains and drawing off in veils over the peaks. The brush we passed through was full of glistening spiders' webs, and in the open the grass was overlaid with disks of cobweb, flashing rainbow colors in the sun.

As we loped gayly along down the curving road, a startled quail would call out, "Who-are-you′-ah? who-are-you′-ah?" and another would cry "quit" in sharp warning tones; while a pair would scud across the road like little hens, ahead of the horse; or perhaps a covey would start up and whirr over the hillside. The sound of Canello's flying hoofs would often rouse a long-eared jack-rabbit, who with long leaps would go bounding over the flowers, to disappear in the brush.

The narrow road wound through the dense bushy undergrowth known as 'chaparral,' and as Canello galloped round the sharp curves I had to bend low under the sweeping branches, keeping alert for birds and animals, as well as Mexicans and Indians that we might meet.

This corner of the valley was the mouth of Twin Oaks Canyon, and was a forest of brush, alive with birds, and visited only by the children whose small schoolhouse stood beside the giant twin oak from which the valley post-office was named. Flocks of migrating warblers were always to be found here; flycatchers shot out at passing insects; chewinks scratched among the dead leàves and flew up to sing on the branches; insistent vireos cried *tu-whip' tu-whip' tu-whip' tu-wee'-ah*, coming out in sight for a moment only to go hunting back into the impenetrable chaparral; lazuli buntings sang their musical round; blue jays — blue squawkers, as they are here called — went screaming harshly through the thicket; and the clear ringing voice of the wren-tit ran down the scale, now in the brush, now echoing from the bowlder-strewn hills above. But the king of the chaparral was the great brown thrasher. His loud rollicking song and careless independent ways, so suggestive of his cousin, the mocking-bird, made him always a marked figure.

There was one dense corner of the thicket where a thrasher lived, and I used to urge Canello through the tangle almost every morning for the pleasure of sharing his good spirits. He was not hard to find, big brown bird that he was, standing on the top of a bush as he shouted out boisterously, *kick'-it-now, kick'-it-now, shut'-up shut'-up, dor'-a-thy dor'-a-thy*; or, calling a halt in his mad rhapsody, slowly drawled out, *whoa'-now, whoa'-now*. After listening to such a tirade as this, it was pleasant to come to an opening in the brush and find a band of gentle yellow-birds leaning over the blossoms of the white forget-me-nots.

There were a great many hummingbirds in the chaparral, and at a certain point on the road I was several times attacked by one of the pugnacious little warriors. I suppose we were treading too near his nest, though I was not keen-eyed enough to find it. From high in the air, he would come with a whirr, swooping down so close over our heads that Canello started uneasily and wanted to get out of the way. Down over our heads, and then high up in the air, he would swing back and forth in an arc. One day he must have shot at us half a dozen times, and another day, over a spot in the brush near us, — probably where the nest was, — he did the same thing a dozen times in quick succession.

In the midst of the brush corner were a number of pretty round oaks, in one of which the warblers gathered. My favorite tree was in blossom and alive with buzzing insects, which may have accounted for the presence of the warblers. While I sat in the saddle watching the dainty birds decked out in black and gold, Canello rested his nose in the cleft of the tree, quite unmindful of the busy warblers that flitted about the branches, darting up for insects or chasing down by his nose after falling millers.

One morning the ranchman's little girl rode

over to school behind me on Canello, pillion fashion. As we pushed through the brush and into the opening by the schoolhouse, scattered over the grass sat a flock of handsome black-headed grosbeaks, the western representative of the eastern rose-breast, looking, in the sun, almost as red

Black-headed Grosbeak.
(One half natural size.)

Rose-breasted Grosbeak.
(One half natural size.)

as robins. They had probably come from the south the night before. As we watched, they dispersed and sang sweetly in the oaks and brush.

In the giant twin oak under whose shadow the the little schoolhouse stood was an owl's nest. When I stopped under it, nothing was to be seen but the tips of the ears of the brooding bird. But when I tried to hoot after the manner of owls, the angry old crone rose up on her feet above the nest till I could see her round yellow eyes and the full length of her long ears. She snapped her bill fiercely, bristled up, puffing out her feathers and shaking them at us threateningly. Poor old bird! I was amused at her performances, but one of her little birds lay dead at the foot of the tree, and I trembled for the others, for the school-children were near neighbors. Surely the old bird needed all her devices to protect her young. One day I saw on one side of the nest, below the big ears of the mother, the round head of a nestling.

It was pleasant to leave the road to ride out under the oaks along the way. There was always the delightful feeling that one might see a new bird or find some little friend just gone to housekeeping. One morning I discovered a bit of a wren under an oak with building material in her bill. She flew down to a box that lay under the tree and I dismounted to investigate. A tin can lay on its side in the box, and a few twigs and yellowish brown oak leaves were scattered about in a casual way, but the rusted lid of the can was half turned back, and well out of sight in the inside was a pretty round nest with one egg in it. I was delighted, — such an appropriate place for a wren's nest, — and sat down for her to come back. She was startled to find me there, and stopped on the edge of the board when just ready to jump down. She would have made a pretty picture as she stood hesitating, with her tail over her back, for the sun lit up her gray breast till it almost glistened and warmed her pretty brown head as she looked wistfully down at the box. After twisting and turning she went off to think the matter over, and, encouraged perhaps by my whistle, came back and hopped down into the little nest.

Two weeks later I was much grieved to find that the nest had been broken up. A horse had been staked under the tree, but he could not have done the mischief; for while the eggs were there, the nest itself was all jumbled up in the mouth of the can. I could not get it out of my mind for days. You become so much interested in the families you are watching that you feel as if their troubles were yours, and are haunted by the fear that they will think you have something to do with their accidents. They had taken me on probation at first, and at last had come to trust me — and then to imagine that I could deceive them and do the harm myself!

When Canello and I left the brushy side of the canyon and started across the valley, the pretty little horned larks, whose reddish backs matched the color of the road, would run on ahead of us, or let the horses come within a few feet of them, squatting down ready to start, but not taking wing till it seemed as if they would get stepped on. Sometimes one sat on a stone by the roadside, so busy singing its thin chattering song that it only flitted on to the next stone as we came up; for it never seemed to occur to the trustful birds that passers-by might harm them.

One of our most interesting birds nested in

holes in the open uncultivated fields down the valley, — the burrowing owl, known popularly, though falsely, as the bird who shares its nest with prairie dogs and rattlesnakes. Though they do not share their quarters with their neighbors, they have large families of their own. We once passed a burrow around which nine owls were sitting. The children of the ranchman called the birds the 'how-do-you-do owls,' from the way they bow their heads as people pass. The owls believe in facing the enemy, and the Mexicans say they will twist their heads off if you go round them times enough.

One of our neighbors milked his cows out in a field where the burrowing owls had a nest, and he told me that his collie had nightly battles with the birds. I rode down one evening to see the droll performance, and getting there ahead of the milkers found the bare knoll of the pasture peopled with ground squirrels and owls. The squirrels sat with heads sticking out of their holes, or else stood up outside on their hind legs, with the sun on their light breasts, looking, as Mr. Roosevelt says, like 'picket pins.' The little old yellowish owls who matched the color of the pasture sat on the fence posts, while the darker colored young ones sat close by their holes, matching the color of the earth they lived in. As I watched, one of the old birds flew down to feed its young. A comical little fellow ran up to meet his parent and then scudded back to the nest hole, keeping low to the ground as if afraid of being seen, or of disobeying his mother's commands. When the ranchman came with his cows the small owls ducked down into their burrows out of sight.

Romulus, the collie, went up to the burrows and the old owls came swooping over his back screaming shrilly — the milkers told me that they often struck him so violently they nipped more than his hair! When the owls flew at him, Romulus would jump up into the air at them, and when they had settled back on the fence posts he would run up and start them off again. The performance had been repeated every night through the nesting season, and was getting to be rather an old story now, at least to Romulus. The ranchman had to urge him on for my benefit, and the owls acted as if they rather enjoyed the sport, though with them there was always the possibility that a reckless nestling might pop up its head from the ground at the wrong moment and come to grief. It would be interesting to know if the owls were really disturbed enough to move their nest another year.

When Canello and I faced home on our daily circuit of the valley, we often found the vineyard well peopled. In April, when it was being cultivated, there was a busy scene. All the blackbirds of the neighborhood — both Brewer's and redwings — assembled to pick up grubs from the soft earth. A squad of them followed close at the plowman's heels, others flew up before his horse, while those that lagged behind in their hunt were constantly flying ahead to catch up, and those that had eaten all they could sat around on the neighboring grape-vines. The ranchman's son told me that when he was plowing and the blackbirds were following him, two or three 'bee-birds,' as they call the Arkansas and Cassin's flycatchers, would take up positions on stakes

In Hot Pursuit.
(Brewer's Blackbird and Bee-birds.)

overlooking the flock; and when one of the blackbirds got a worm, would fly down and chase after him till they got it away, regularly making their living from the blackbirds, as the eagles do from the fish hawks.

One day in riding by the vineyard, to my surprise and delight I saw one of the handsome yellow-headed blackbirds sitting with dignity on a grape-vine. Although his fellows often flock with redwings, this bird did not deign to follow the cultivator with the others, but flew off and away while I was watching, showing his striking white shoulder patches as he went. The distinguished birds were sometimes seen assembled farther down the valley; and I once had a rare pleasure in seeing a company of them perched high on the blooming mustard.

The son of the ranchman told me an interesting thing about the ordinary blackbirds. He said he had seen a flock of perhaps five hundred fly down toward a band of grazing sheep, and all but a few of the birds light on the backs of sheep. The animals did not seem to mind, and the birds flew from one to another and roosted and rode to their heart's content. They would drop to the ground, but if anything startled them, fly back to their sheep again. Sometimes he had seen a few of the blackbirds picking out wool for their nests by bracing themselves on the backs of the sheep, and pulling where the wool was loose. He had also seen the birds ride hogs, cattle, and horses; but he said the horses usually switched them off with their tails.

On our way home we passed a small pond made by the spring rains. Since it was the only body of water for miles around, it was especially refreshing to us, and was the rendezvous of all our feathered neighbors — how they must have wished it would last all through the hot summer months! As I rode through the long grass on the edge of the pond, dark water snakes often wriggled away from under Canello's feet; but he evidently knew they were harmless, for he paid no attention to them, though he was mortally afraid of rattlers. I did not like the feeling that any snake, however innocent, was under my feet, so would pull him up out of the grass onto a flat rock overlooking the pond.

In the fresh part of the morning, before the fog had entirely melted away, the round pool at our feet mirrored the blue sky and the small white clouds. If a breath of wind ruffled the water into lines, in a moment more it was sparkling. Along the margin of the water was a border of wild flowers, pink, purple, and gold; on one side stood a group of sycamores, their twisted trunks white in the morning sun and their branches full of singing birds; while away to the south a line of dark blue undulating hills was crowned by the peak from which we had looked off on the mountains of Mexico. The air was ringing with songs, the sycamores were noisy with the chatter of blackbirds and bee-birds, and the bushes were full of sparrows.

There was an elder on the edge of the pond, and the bathers flew to this and then flitted down to the water; and when they flew up afterwards, lighted there to whip the water out of their feathers and sun themselves before flying off. I never tired watching the little bathers on the beach. One morning a pipit came tipping and tilting along the sand, peeping in its wild, sad way. Another time a rosy-breasted linnet stepped to the edge of the pond and dipped down daintily where the water glistened in the sunshine, sending a delicate circle rippling off from its own shadow. Then the handsome white and golden-crowned sparrows came and bathed in adjoining pools. When one set of birds had flown off to dry their feathers, others took their places. A pair of blackbirds walked down the sand beach, but acted absurdly, as if they did not know what to do in water — it was a wonder any of the birds did in dry California! Two pieces of wood lay in the shallows, and the blackbirds flew to them and began to promenade. The female tilted her tail as if the sight of herself in the pond made her dizzy, but the male finally edged down gingerly and took a dip or two with his bill, after which both flew off.

On the mud flats on one side of the pond, bee-birds were busy flycatching, perching on sticks near the ground and making short sallies over the flat. Turtle doves flew swiftly past, and high over head hawks and buzzards circled and let themselves be borne by the wind.

Swallows came to the pond to get mud for their nests. A long line of them would light on the edge of the water, and then, as if afraid of wetting their feet, would hold themselves up by fluttering their long pointed wings. They would get a little mud, take a turn in the air, and come back for more, to make enough to pay them for their long journeys from their nests. Sometimes they would skim over the pond without touching the surface at all, or merely dip in lightly for a drink in passing; at others they would take a flying plunge with an audible splash. Now and then great flocks of them could be seen circling around high up against a background of cloud's and blue sky.

One day I had a genuine excitement in seeing a snow-white egret perched on a bush by the water. I rode home full of the beautiful sight, but alas, my story was the signal for the ranchman's son to seize his gun and rush after the bird. Fortunately he did not find him, although he did shoot a green heron; but it was probably a short reprieve for the poor hunted creature.

Canello was so afraid of miring in the soft ground that it was hard to get him across some places that seemed quite innocent. He would test the suspicious ground as carefully as a woman, one foot at a time; and if he judged it dangerous, would take the bits, turn around and march off in the opposite direction. I tried to force him over at first, but had an experience one day that made me quite ready to take all suggestions in such matters. This time he was deceived himself. We were on our homeward beat, off in the brush beyond the vineyard. I was watching for chewinks. We came to what looked like an old road grown up with soft green grass, and it was so fresh and tender I let Canello graze along at will; while keeping my eyes on the brush for chewinks. Suddenly Canello pricked up his ears and raised his head with a look of terror. Rattlesnakes or miring — it was surely one or the other! When I felt myself sinking, I knew which. I gave the horse a cut with the quirt to make him spring off the boggy ground, and looked off over his side to see how far down he was likely to go, but found myself going down backwards so fast I had to cling to the pommel. I lashed Canello to urge him out, and he struggled desperately, but it was no use. We were sinking in deeper and deeper, and I had to get off to relieve him of my weight. By this time his long legs had sunk in up to his body. On touching the ground I had a horrible moment thinking it might not hold me; but it bore well. Seizing the bridle with one hand and swinging the quirt with the other, I shouted encouragement to Canello, and, straining and struggling, he finally wrenched himself out and stepped on *terra firma* — I never appreciated the force of that expression before! The poor horse was trembling and exhausted when I led him up to high ground to remount, and neither of us had any desire to explore boggy lands after that.

On our morning round, Canello and I attended strictly to business,—he to grazing, I to observing; but on our afternoon rides I, at least, felt that we might pay a little more heed to the beauties of the valley and the joys of horsebacking. Sometimes we would be overtaken by the night fog. One moment the mustard would be all aglow with sunshine; at the next, a sullen bank of gray fog would have risen over the mountain, obscuring the sun which had warmed us and lighted the mustard; and in a few moments it would be so cold and damp that I would urge Canello into a lope to warm our blood as we hurried home.

II.

THE LITTLE LOVER.

The Little Lover.
(Western House Wren.)

ON my second visit to California, I spent the winter in the Santa Clara valley, riding among the foothills of the Santa Cruz Mountains, where flocks of Oregon robins were resting from the labors of the summer and passing the time until they could fly home again; but when the first spring wild flowers bloomed on the hills I shipped my little roan mustang by steamer from San Francisco to San Diego, and hurried south to meet him and spend the nesting season in the little valley of the Coast Mountains which, five years before, had proved such an ideal place to study birds.

I went down early in March, to be sure to be in time for the nesting season; but spring was so late that by the last of April hardly a nest had been built, and it seemed as if the birds were never coming back. The weather was gloomy and the prospect for the spring's work looked discouraging, when one morning I rode over to the line of oaks and sycamores at the mouth of Ughland canyon I had not visited before. In this dry, treeless region of southern California only a little water is needed to cover the bare valley bottoms with verdure. The rushing streams that flow down the canyons after the winter rains fill their mouths with rich groves of brush, oaks and sycamores; while lines of trees border the streams as far as they extend down the valleys. Before the streams go far, the thirsty soil drinks them up, leaving only dry beds of sand bordered by trees, until the rains of the following winter. In April, the water in this particular canyon mouth had already disappeared, and the wide sand bed under the trees alone remained to tell of the short-lived stream. But the resulting verdure was enough to attract the birds. Apparently a party of travelers had just arrived. The brush and trees were full of song — yellowbirds, linnets, chewinks, doves, wrens, and, best of all, a song sparrow, — bless his heart! — singing as if he were on a bush in New York state. It was more cheering than anything I had heard in California.

When able to listen to something besides song sparrows, I realized that from the trees in front of me was coming the rippling merry song of a wren. Wrens are always interesting, — droll, individual little scraps, — and having found their nests in sycamore holes before, I let my horse, Mountain Billy, graze nearer to the tree from which the sound came. Before long the small brown pair flew away together across the oat field that spread out from the mouth of the canyon. While they were gone, I took the opportunity to inspect the tree, and found a large hole with twigs sticking out suggestively. Presently, back flew one of the wrens with more building material. But this line of sycamores was off from the highway, and the bird was not used to prying equestrians; so when she found Mountain Billy and me planted in front of her door, she doubted the wisdom of showing us that it was her door. Chattering nervously, she would back and fill, flying all but to the door and then flitting off again. She could not make up her mind to go inside. But soon her mate came and — unmindful of visitors, ardent little lover that he was — sang to her so gayly that it put her in heart; and before I knew it she had slipped into the tree.

Here was a nest, at last, right over my eye. To encourage myself while waiting for something to happen, I began a list with the heading NESTS, when something caught my eye overhead, and glancing up, behold, a goldfinch walked down a branch and seated herself in a round cup! A few moments later — buzz — whirr — a hummingbird flew to a nest among the brown leaves of one of the low-hanging oak sprays not ten feet away! I simply stared with delight and astonishment. No need of a list for encouragement now. From Billy's back I could look down into the little cup, which seemed the tiniest in the world. Forgetting the little lover and his mate, I sat still and watched this small household.

The young were out of the eggs, though not much more, and their mother sat on the edge of the nest feeding them. She curved her neck over till her long bill stood up perpendicularly, when she put it gently into the gaping bills of her young; the smallest of bills, not more than an eighth of an inch long, I should judge. I never saw hummingbirds fed so gently. Probably the small bills and throats were so delicate the mother was afraid they would not bear the usual jabbing and pumping.

When the little ones were fed, the old bird got down in the nest, fluffing her feathers about her in a pretty motherly way and settling herself comfortably to rest, apparently ignoring the fact that Billy was grazing close beside her. She may have had her qualms, but no mother bird would leave her tender young uncovered on such a cold morning.

While she was on the nest, there was an approaching whirr, followed by a retreating buzz — had the father bird started to come to the nest and fled at sight of me? Remembering the evidence Bradford Torrey collected to prove that the male bird is rarely seen at the nest, I wondered if his absence might be explained by his usually noisy flight, for it would attract the notice of man or beast.

Two days later I carefully touched the tip of my finger to the back of one of the tiny hummingbirds, — it was very skinny, I regret to state, — and at my touch the little thing opened its wee bill for food. That day the mother fed the birds in the regulation way, when we were only four feet distant. I was near enough to see all the horrors of the performance. She thrust her bill down their throats till I felt like crying out, "For mercy's sake, forbear!" She plunged it in up to the very hilt; it seemed as if she must puncture their alimentary canals.

While waiting for the wrens, I buckled Billy's bridle around the sycamore and threw myself down on the warm sand under the beautiful tree. The little horse stood near, outlined against the blue sky, with the sunlight dappling his back, while I looked up into the light green foliage of the white sycamore overhead. There seemed to be a great deal of light stored in these delicate trees. The undersides of the big, soft, white leaves looked like white Canton flannel; the sunlight mottled the whitish bark of the trunks and branches; and a great limb arched above me, making a high vaulted chamber whose skylights showed the deep blue above.

But there were the little lover and his mate, and I must turn my glass on them. She came first, with long streamers hanging from her bill, and at sight of me got so flustered that one of her straws slipped out and went sailing down to the ground. When the pair had gone again, two linnets came along. The female saw the wren's doorway, and being in search of apartments flew up to look at the house. When she came out she and her mate talked it over and, apparently, she told him something that aroused his curiosity — perhaps about the wren's twigs she found inside — for he flew into the dark hole and looked around as she had done. Then both birds went off to inspect other holes in the tree. The master of the wren cottage came back in time to see them on their rounds, and taking up his position in front of his door sang out loudly, with wings hanging and a general air of, "This is *my* house, I'd have you understand!"

When the lord of the manor had flown away, his lady came. I thought perhaps he had told her of the visitors and she had come to see if they had disturbed any of her sticks, for she brought no material. She was afraid to go to the nest in my presence, but flew to a branch near by and leaned down so far it was a wonder she did n't tip over as she stared anxiously at the hole — a bad way to keep a secret, my little lady! I thought. When her merry minstrel came, his song again gave her courage and she flew inside, turning in the doorway, however, to look out at me.

But what with horses grazing under her windows and linnets making free with her nest, the poor wren was unsettled in her mind. Possibly it would be wiser to take out her sticks and build elsewhere. She went about looking at vacant rooms and examined one opening in the side of the trunk where I could see only her profile as she hung out of the hole.

For some time the timid bird would not accept Mountain Billy and me as part of her immediate landscape, and I watched the premises a number of days, getting nothing but my labor for my pains, as far as wrens were concerned.

One day when she did not come, I thought it was a good chance to get a study of the hummingbird's nest; but alas! — the delicate little structure hung torn and dangling from the twig, with nothing to tell what had become of the poor little hummers. I moralized sadly upon the mutability of human affairs as I took the tattered nest and tied it up in a corner of my handkerchief; for it was all that was left of the little home built with such exquisite care and brooded over so tenderly.

The yellowbird's nest came to an untimely end, too, although its start was such a bright one. It was a disappointment, for the goldfinches are such trustful birds and so affectionate and tender in their family relations that they always win one's warm interest. At first, when this mother bird went to the nest, her mate stationed himself on the nest tree, leaning over and looking down anxiously at Billy and me; but before their home was broken up the watchful guardian fed his pretty mate at her brooding when we were below.

We had a great many visitors while waiting for the wrens: neighbors came to sit in our green shade, young housekeepers came looking for rooms to rent, and old birds who were leading around their noisy families came to dine with us. Once a pair of flickers started to light in the tree, but they gave a glance over the shoulder at me and fled. Later I found their secret — down inside an old charred stump up the canyon. Occasionally I got sight of gay liveries in the green sycamore tops. A Louisiana tanager in his coat of many colors stopped one day, and another time, when looking up for dull green vireos, my eye was startled by a flaming golden oriole. The color was a keen pleasure. Lazuli buntings, relatives of our eastern indigo-bird, sang so much within hearing that I felt sure they were nesting in the weeds outside the line of sycamores — I did find a pair building in the malvas beyond; a pair of bush-tits, cousins of the chickadees, came with one of their big families ; California towhees often appeared sitting quietly on the branches ; linnets were always stopping to discuss something in their emphatic way; clamorous blue jays rushed in and set the small birds in a panic, but seeing me quickly took themselves off ; and a pair of wary woodpeckers hunted over the sycamore trunks and worked so cautiously that they had finished excavating a nest only just out of my sight on the other side of the wren tree trunk before I seriously suspected them of domestic intentions.

One day, when watching at the tree, a great brown and black lizard that the children of the valley call the 'Jerusalem overtaker' came worming down the side of an oak that I often leaned against. The rough bark seemed such a help to it that I imagined the wrens had done wisely in choosing a smooth sycamore to build in. I looked narrowly at their nest hole with the thought in mind and saw that the birds had another point of vantage in the way the trunk bulged at the hole — it did not seem as if a large lizard could work itself up the smooth slippery rounding surface, however much given to eggs for breakfast. But in the West Indies lizards walk freely up and down the marble slabs, so it is dangerous to say what they cannot do.

Billy had a surprise one day greater than mine

over the lizard. He was grazing quietly near where I sat under the wren tree, when he suddenly threw up his head. His ears pointed forward, his eyes grew excited, and as he gazed his head rose higher and higher. I jumped from the ground and put my hand on the pommel ready to spring into the saddle. As I did so, across the field I caught a glimpse of a great fawn-colored animal with a white tip to its tail, bounding through the brush — a deer! Then I heard voices through the trees and saw the red shawl of a woman in a wagon rumbling up the road the deer must have crossed.

When Mountain Billy and I pulled ourselves together and started after the deer, the poor horse was so unstrung he made snakes of all the sticks he saw and shied at all imaginable bugaboos along the way. We were too late to see the deer again, but found the marks of its hoofs where it had jumped a ditch and sunk so deep in the fine sand on the other side that it had to take a great leap to recover itself.

The sight of the deer made Billy as nervous as a witch for days. Every time we went to visit the wrens he would stand with eyes glued to the spot where it had appeared, and when a jack-rabbit came out of the brush with his long ears up, Billy started as if he thought it would devour him. I was perplexed by his nervousness at first, but after much pondering reasoned it out, to my own satisfaction at least. His name was Mountain Billy, and in the days when he had been a wayward bucking mustang he lived in the Sierra. Now, even in the hills surrounding our valley, colts were killed by mountain lions. How much more in the Sierra. Mountain lions are large fawn-colored animals: that was it: Mountain Billy was suffering from an acute attack of association of ideas. The sight of the deer had awakened memories of the nightmare of his colthood days.

We made frequent visits to the wren tree, and both my nervous little horse and I had a start one morning, for as we rode in, a covey of quail flew up with a whirr from under the tree in front of us.

When the wren had become reconciled to us she worked rapidly, flying back and forth with material, followed by her mate, who sang while she was on the nest and chased away with her afterwards. Often when she appeared in the doorway ready to go, his song, which had been just a merry round before, at sight of her would suddenly change to a most ecstatic love song. He would sit with drooping tail, his wings sometimes shaking at his sides, at others raised till they almost met over his back, trembling with the excitement of his joy. This peculiar tremulous motion of the wings was marked in both wrens; their emotions seemed too large for their small bodies.

I found the wrens building, the last of April.

The third week in May the little lover was singing as hard as ever. I wrote in my note-book — "Wrens do not take life with proper seriousness, their duties certainly do not tie them down." When the eggs were in the nest, if her mate sang at her door, the mother bird would fly out to him and away they would go together; for it never seemed to occur to the care-free lover that he might brood the eggs in her absence.

When the young hatched, however, affairs took a more serious turn. Mother wren at least was kept busy looking for spiders, and later, when both were working together, if not hunting among the green treetops, the pretty little brown birds often flew to the ground and ran about under the weeds to search for insects. Once when the mother bird had flown up with her bill full, she suddenly stopped at the twig in front of the nest, looking down, her tail over her back wren fashion, the sun on her brown sides, and her bill bristling with spiders' legs.

On June 7 I noticed a remarkable thing. For more than five weeks, all through the building and brooding, the little lover had been acting as if on his honeymoon — as if the nest were a joke and there were nothing for him to do in the world but sing and make love to his pretty mate — as if life were all 'a-courtin'.' On this day he first came to the tree with food, sang out for his spouse, gave her the morsel, and flew off. Later in the morning he brought food and his mate carried it to the young. But afterwards, when she started to take a morsel from him, behold! he — the gay, frivolous little beau, the minstrel lover — actually acted as if he did n't want to give it up, as if he wanted to feed his own little birds himself. With wings trembling at his sides he turned his back on his mate and started to walk down the branch away from her! But he was too fond of her to even seem to refuse her anything, and so, coming back, gave her the morsel. She probably divined his thought, and, let us hope, was glad to have him show an interest in his children at last; at all events, when he came again with food and clung to the tip of a drooping twig waiting

A Trying Moment.

although she first lit above him and came down toward him with bill wide open and wings fluttering in the pretty, helpless, coquettish way female birds often tease to be fed; suddenly, as if remembering, she flew off, and — he went in to the nest himself! It was a conquest; the little lover was not altogether lacking in the paternal instinct after all! I looked at him with new respect.

On June 12 I wrote: "The wrens seem to have settled down to business." It was delighful to find the small father actually taking turns feeding the young. I saw him feed his mate only once or twice, and noticed much less of the quivering wings, though after leaving the nest he would sometimes light on a branch and move them tremulously at his sides for a moment. June 15 I wrote: "The birds are feeding rapidly to-day. I hear very little song from the male; probably he has all he can attend to. I'd like to know how many young ones there are in that hole." At all events, the voices of the young were getting stronger and more insistent, and it is no bagatelle to keep half a dozen gaping mouths full of spiders, as any mother bird can tell. This particular mother wren, however, seemed to enjoy her cares. She often called to the young from a branch in front of the nest before going in, and stopped to call back to them with a motherly-sounding *krup-up-up* as she stood in the entrance on leaving.

One day as one of the old birds stood in the doorway its mate flew into the nest right over its head. The astonished doorkeeper was so startled that it took to its wings.

Before this, in watching the wrens, I had looked off across a sunny field of golden oats, against the background of blue hills. On June 14, when I went to the nest, the mowers had been at work around the sycamores and the oat-field was full of cocks. Just as the wren was most anxious for peace and quietness, for a safe world into which to launch her brood, up came this rout of haymakers with all their clattering machines, laying low the meadows to her very door.

No wonder the little bird met me with nerves on edge. When the eggs had first hatched, she had objected to me, but mildly. To be sure, once when she found me staring she flew away over my head, scolding as much as to say, "Stop looking at my little birds," and finding me there when she came back, shook her wings at her sides and scolded hard, though her bill was full; but still her disapproval did not trouble me; it was too sociable. But now, for some time, affected by the shadow of coming events, she had been growing more and more fidgety under my gaze, darting inside, then whisking back to the door to look at me, in again to her brood and out to me, over and over like a flash — or, like a poor little troubled mother wren, distracted lest her unruly youngsters should pop out of the hole in the tree trunk when I was below to catch them.

On this day, when the wren came up from the dark nest pocket and found me below, she called back to her little ones in such distress that I felt reproached. By gazing fixedly through my glass into the dark hole I could see the head of a sprightly nestling pop up and turn alertly from side to side as if returning my inspection. The old wren's calls made me think of a human mother who can no longer control her big wayward offspring and has to entreat them to do as she bids. It was as if she said, "Oh, *do* be good children, *do* keep still; *do* put your heads back; you *naughty* children, you *must* do as I tell you!"

On June 16, six weeks after I had found the birds building, I wrote in my note-book: "I am astonished every morning when I come and find the wrens still here, but perhaps it's easier feeding them in one spot than it would be chasing around after them in half a dozen different places."

The young were chattering inside the nest. They all talked at once as children will, but one small voice assumed the tones of the mother; probably the oldest brother speaking with the air of authority featherless children sometimes assume with the weaker members of the family. When a parent came, I saw the big brother's head pop up from behind the wall, — the nest was in a pocket below, — and by the time the old bird got there with food the big throat blocked the way for the little ones down behind. Sometimes I could see a flutter of small wings and tails when the birds were being fed.

As nothing happened, I went off to watch another nest, but in an hour was back to make sure of seeing the small wrens when they left the nest. A loud continuous scolding met me on approaching, and one of the old wrens, with bill full of insects, flew — not up to the nest — but down in among the weeds! In less than an hour that whole brood of wrens had flown, and were three or four rods away in the high weeds — safe! I was taken aback. They had stolen a march on me. Surely I had not been treated as was fit and proper, being one of the family!

It was amusing to see the young ones fly. They whirled away on their wings as if they had been flitting around in the big world always; but their stubby tails sadly interfered with their progress, and they came to earth before they meant.

Weak cries came from the young hidden in the weeds. They could fly, but it was different from being safe inside a tree trunk! I hardly recognized their weak appealing voices, after the stentorian tones that had issued from the old nest.

The weeds were a most admirable cover, and the dead stalks sticking up through them served as sentry posts, from which the old birds scolded me when I followed too close on their heels. The youngsters sometimes appeared on the stalks, and looked very pert on their long legs with their short tails cocked over their backs.

In the afternoon I went again to see the little family to which I had become so much attached and which were now slipping away from me. They had been led farther up the canyon, where, at a turn in the dry bed of the stream, the thick cover of weeds was still more protected by brush and overhanging trees, and the whole thicket was warmed by the afternoon sunshine. The old birds were busily flying back and forth feeding their invisible young. They scolded me as they flew past, but kept right on with their work.

There was little use trying to keep track of the brood after that, and I thought I had given them up quite philosophically, reflecting that it was pleasant to leave them in such a sunny protected place. Still, day after day in riding along the line of sycamores on my way to other nests, it gave me a pang of loneliness to pass the old deserted wren tree where I had spent so many happy hours; and though the sycamores were silent, I could always hear and see the little lover singing to his pretty mate.

III.

LIKE A THIEF IN THE NIGHT.

When watching the little lover and his brood, I heard familiar voices farther down the line of oaks, voices of little friends I had made on my first visit to California, and had always remembered with lively interest as the jauntiest, most individual bits of humanity I had ever known in feathers. So, when Mountain Billy and I could be spared by the other bird families we were watching, we set out to hunt up the little bluish gray western gnatcatchers.

The (sand) stream that widened under the wren's sycamores narrowed up the canyon to a — dry ditch, I should say, if it were not disrespectful to speak that way of a channel that once a year carries a torrent which excavates canals in the meadows. Billy and I started up this sand ditch, so narrow between its weed-grown banks that there was barely room for us, and so arched over in places by chaparral that we could get through only when Billy put down his ears and I bowed low on the saddle.

We had not gone far before we heard the gnatcatchers, bluish gray mites with heads that are

Nest of Western Gnatcatcher.

(From a photograph.)

always cocked on one side or the other to look down at something, and long tails that are always flipping about as their owners flaunt gayly through the bushes. At sound of their voices I pulled Billy up out of the ditch, and, slipping from his back, sat down on the ground to wait for the birds. Eureka! there, in a slender young oak on the edge of the stream not a rod away, one of the pair was gliding off its nest, a beautiful lichen-covered, compact little structure such as I had admired years before. I was jubilant. What a relief! I had fully expected it to be inside the dense brush, where no mortal could tell what was going on; and here it was out in the plain light of day. What a delightful time I should have watching it! Before leaving the spot, in imagination I had followed the brood out into the world and filled a note-book with the quaint airs and graces of the piquant pair.

When insinuating yourself into the secrets of the bird world, it is not well to be too obtrusive at first: it is a mistake to spend the day when you make your first call; so contenting myself with thinking of the morrow, and fixing the small oak in my memory, I took myself off before the blue-gray should tell on me to her mate. As I rose to go, a dove flew out of the oak — she had been brooding right over my head. Another nest, and a mourning dove's, one of the most gentle and winning of birds! Surely my good star was in the ascendent!

The next day, forgetful of this second nest, I rode Billy right up under the oak, and was startled to find the pretty dove sitting quietly over our heads, looking down at us out of her gentle eyes. It was a pleasant surprise. She let me talk to her, but when I had dismounted Billy tramped around so uneasily that the saddle caught in the oak branches and scared the poor bird away. I had hardly seated myself when the jaunty little gnatcatcher came flying over and lit in an upper branch of the tree. What a contrast she was to the quiet dove! With many flirts of the tail she hopped down to the nest, jumping from branch to branch as if tripping down a pair of stairs. When she dropped into her deep cup her small head stuck up over one edge, her long tail pointed over the other.[1]

I looked away a moment, and on glancing back found the nest empty. On the instant, however, came the sound of my small friend's voice. Such a talkative little person! — not one of your creep-in-and-out-of-the-nest-without-anybody's-knowing-it kind of a bird, not she! Her remarks sounded as if made over my head, and when Billy stamped about the brush and rapped the saddle trying to switch off flies, I imagined guiltily that they were addressed to me; but while I wondered if she would keep away all the rest of the morning because she had discovered me, back she came, talking to herself in complaining tones and whipping her tail impatiently, even after she stood on the edge of the nest, evidently absorbed in her own affairs, quite to the exclusion of the person down in the brush who thought herself so important!

My doves were attending to me, however, altogether too much. The brooding bird was anxious to go to her nest. After flying out where she

[1] As this little pair dressed like twins, I could only infer which was which from the song and the actions of the two, which were quite distinct.

could see me, she whizzed toward it; but, fearful, hesitated and talked it over with her mate — both birds cooed with inflated breaths. After that the branches rattled overhead, but even then, though my back was turned, the timid bird dared not stay. She must make another inspection. From an opposite oak she peered through the branches, moving her head excitedly, and calling out her impressions to her mate. Meanwhile, he had flown down the sand stream and called back quite calmly. I, also, cooed reassuringly to her, and soon she quieted down and began to plume her feathers on the sunny branch. As the gnatcatchers did not honor us with their attention even when Billy stalked around in plain sight, I moved a little closer to their nest to give the dove more freedom; and soon the gentle bird slipped back to her brooding.

Before leaving I went to see the dove in the oak, and spoke caressingly to her, admiring her soft dove-colored feathers and shining iridescent neck. She was on her own ground there, and felt that she could safely be friends, so she only winked in the sun, paying no heed to her mate when he called warningly. It was especially pleasant to watch this reserved lady-like bird, after the flippant tell-all-you-know little gnat.

On going away, Billy and I took a run up the canyon. Billy was in high spirits, and went racing up the narrow road, winding and turning through the chaparral, brushing me against the the stiff scrub oak and loping under low branches so fast that the sharp leaves snapped back, stinging my cheeks. We had a gay ride, with a spice of excitement thrown in; for on our way home, in the thick dust across our path, besides the pretty quail tracks that made wall-paper patterns on the road, were the straight trails of gopher snakes, and the scalloped one of a rattlesnake we had been just too late to meet.

At our next session with the blue-grays, when she was on the nest, her mate came back to relieve her and cried in his quick cheerful way, "Here I am, here I am!" Either she was taking a nap or did n't want to stir, for she did n't budge till he called insistently, "*Here* I am, *here* I am!" Then he hopped down in her place, and raising his head above the nest, remarked again, as if commenting upon the new situation, "Here I am!"

It was quite a different matter when she came back to work. She only called "hello," not even hinting that he should make way for her, but he hopped off at the first sound of her voice, flying away promptly to another tree and calling back like a gleeful boy let out of school, "Here I am!"

She was no more eager to go to the nest than he, however, and once when she came flirting leisurely along from twig to twig, she stopped a long time on the edge of the nest and leaned over, presumably to arrange the eggs; perhaps she and her mate had different views as to their proper positions. The next time I visited the gnats, she acted as if she really could not make up her mind to settle down to brooding on such a beautiful morning. The fog had cleared away and the air was fresh and full of life; goldfinches and lazuli buntings were singing merrily, and light-hearted vireos were shouting *chick-a-de-chick'-de-villet'* from the brush. How much pleasanter it would be for such an airy fairy to go off for a race with her mate than to settle down demurely tucked into a cup! "Tsang," she cried impatiently as she flew up to catch a fly. She flirted about the branches, whipped up in front of the nest, could n't make up her mind to go in, and flounced off again. But the eggs would get cold if she did n't cover them, so back she came, hopped up on the edge of the nest, and stood twisting and turning, glancing this way and that as though for a fly to chase, till she happened to look down at the eggs; then she whipped her tail, dropped in and — jumped out again!

During the morning when she was away and her mate was waiting for her to come back to 'spell' him, he too got impatient. He hopped out of the nest crying, "Now here I am, quick, come quick!" and as he flew off, sang out in his funny little soliloquizing way, "Well, here I go; here I go!"

His restless spouse had only just settled down when a wren-tit — a wren-like bird with a long tail — flew into a bush near her oak, and she darted out of the nest to snap her bill over his head. I thought it merely an excuse to leave her brooding. Calling out "tsang," she again flew at the brown bird who was hopping around in the bush, so innocently, as I thought. Conqueror for the moment, she flaunted back to the nest, and after much ado finally settled down.

For a time all was quiet. Hearing the low cooing of doves, I went to talk to the pretty bird in the oak, and she let me come near enough to see her bluish bill and quiet eyes. As I returned to the gnatcatchers, a chewink was hoeing in the sand stream. Again the wren-tit approached stealthily. I watched with languid interest till he got to the gnat's tree. The instant he touched foot upon her domain, she dashed down at him, crying loudly and snapping her bill in his face. The brown bird dodged her blows, held his footing in spite of her, and slowly made his way up to the nest. I was astonished and frightened. He leaned over the nest, and — what he actually did I could not see, for by that time the blue-gray's cries had called her mate and they were both screaming and diving down at him as if they would peck his eyes out; and it sounded as if they hit him on the back good and hard.

A peaceful lazuli bunting, hearing the commotion, came to investigate, but when she saw what was happening held back against the side of a twig as though afraid of getting struck, and soon flew off, having no desire to get mixed up in that affray.

When the wren-tit had at last been driven from his position, the gnatcatchers flew up into a tree and, standing near together, talked the matter over excitedly. Then one of them went back to the nest, reached down into it and brought up something that it appeared to be eating. Its mate went to the nest and did the same, after which one of them flew away with a broken eggshell. When the little creatures turned away from the plundered nest they broke out into cries of distress that were pitiful to hear. I felt indignant at the wren-tit. How could a bird with eggs of its own do such a cruel thing? But then, I reflected, we who pretend to be better folks than wren-tits do not always spare our neighbors because of our own troubles. When the poor birds had carried away their broken eggshell, one of them came and tugged at the nest lining till it pulled out a long horse-hair and what looked like a feather, apparently trying to take out everything that the egg had soiled.

When the little housekeeper was working over her nest, a brown towhee flew into the tree. On the instant there was a flash of wings — the gnat was ready for war. But after a fair look at the big peaceful bird, she flew to the next tree without a word — she evidently knew friends from enemies. I never liked the towhee so well before. But though the blue-gray had nothing to say against her neighbor sitting up in the tree if he chose, her nerves were so unstrung that when she lit in the next tree she cried out "tsang" in an overburdened tone. It sounded so unlike the usual cry of the light-hearted bird, it quite made me sad.

Whether the poor little gnatcatchers did not recover from this attack upon their home, and took their nest to pieces to put it up elsewhere, as birds sometimes do; or whether the stealthy wren-tit again crept in like a thief in the night to plunder his neighbor's house, I do not know; but the next time I went to the oak the nest was demolished. It was a sorry ending for what had promised to be such an interesting and happy home.

My poor dove's nest had a tragic end, too. What happened I do not know, but one day the body of a poor little pigeon lay on the ground under the nest. My sympathies went out to both mothers, but especially to the gentle dove, now a mourner, indeed.

IV.

WAS IT A SEQUEL?

After the wren-tit stole in like a thief in the night and broke up the pretty home of the gnat-catchers, I suspected that they took their house down to put it up again in a safer place, and so was constantly on the lookout to find where that safer place was. At last, one day, I heard the welcome sound of their familiar voices, and following their calls finally discovered them flying back and forth to a high branch on an old oak-tree; both little birds working and talking together. Mind, I do not stake my word on this being the same pair of gnats; but the nest followed closely on the heels of the plundered one, which was a point in its favor, and, being anxious to take up the lines with my small friends again, I let myself think they were the birds of the sand ditch nest. It was such a delight to find them that I deserted the nest I had been watching, and went to spend the next morning with my old friends. The tree they had chosen was a high oak in an open space in the brush, and they were building fifteen or twenty feet above the ground — so high that it was necessary to keep an opera-glass focused on the spot to see what was going on at their small cup.

As the birds worked, I was filled with forebodings by seeing a pair of wren-tits on the premises. They went about in the casual indifferent way sad experience had shown might cover a multitude of evil intentions, and which made me suspect and resent their presence. How had they found the poor little gnats? It was not hard to tell. How could they help finding such talkative fly-abouts? But if birds are in danger from all the world, including those who should be their comrades and champions, why should not builders keep as still at the nest as brooding birds, instead of heedlessly giving information to observers that lurk about taking notes for future misdeeds? But then, could gnatcatchers keep still anywhere at any time? No, that was not to be hoped for. I could only watch the little chatterers from hour to hour and be thankful for every day that their home was unmolested.

It was interesting to see how the jaunty indifferent gnats would act when settling down to plain matters of business. Strange to say, they proved to be the most energetic, tireless, and skillful of builders. Their floor had been laid — on the branch — before I arrived on the scene, and they were at work on the walls. The plan seemed to be twofold, to make the walls compact and strong by using only fine bits of material and packing them tightly in together; while at the same time they gave form to the nest and kept it trim and shipshape by moulding inside, and smoothing the rim and outside with neck and bill. Sometimes the bird would smooth the brim as a person sharpens a knife on a whetstone, a stroke one way and then a stroke the other. When the sides were not much above the floor, one bird came with a bit of material which it proceeded to drill into the body of the wall. It leaned over and threw its whole weight on it, almost going head first out of the nest, and had to flutter its wings to recover itself. The birds usually got inside to build, but there was a twig beside the nest that served for scaffolding, and they sometimes stood on that to work at the outside.

At first they seemed to take turns at building, working rapidly and changing places quite regularly; but one morning when seated under the oak I saw that things were not as they had been. Perhaps a difference of opinion had arisen on architectural points, and Mrs. Gnatcatcher had taken matters into her own hands. At all events, this is what happened: instead of rapid changes of place, when one of the gnats was at work its mate flew up and started to go to the nest, hesitated, and backed away; then unwilling to give up having a finger in the pie, advanced again. This was kept up till the little bird put its pride in its pocket, and gently gave over its cherished bit of material to its mate at the nest!

Now as these gnatcatchers had the bad taste to dress so nearly alike that I could not tell them apart, I was left to my own surmises as to which took the material. Still, who could it have been but Mrs. Gnat? Would she give over the house to Mr. Gnat at this critical moment? She doubtless wanted to decorate as she went along, and men are n't supposed to know anything about such trivial matters! On the other hand, it might easily be he, for, supposing he had come of a family of superior builders, surely he would want to see to the laying of substantial walls; and unquestionably a good wall was the important part of this nest. Alas! it was a clear case of "The Lady or the Tiger." To complicate matters, the birds worked so fast, so high over my head, and so hidden by the leaves, that I had much ado to keep track of their exchanges at all. If I could only catch them and tie a pink ribbon around one of their necks! — then, at least, I would know which was doing what, or if it was doing what it had n't done before! It is inconsiderate enough of birds to wear the same kind of clothes, but to talk alike too, when hidden by the leaves —that, indeed, is a straw to break the camel's back. If small gray gnatcatchers up in the treetops had only been big black magpies low in the brush, my testimony regarding their performances might be of more value; but then, the magpies of my acquaintance were so shy they would have none of me; so although life and field work are full of disappointments, they are also full of compensations.

Not being able to do anything better with the gnat problems, I guessed at which was which — when I saw No. 2 go to the nest and No. 1 reluctantly make way as if not wanting No. 2 to meddle, I drew my own conclusions, although they were not scientifically final. I did see one thing that was satisfactory, as far as it went. One of the birds came with big tufts of stiff moss sticking out from either side of its bill like great mustachios, and going up to the nest, handed them to its mate — actually something big enough for a person to see, once! Whatever had been the birds' first feeling as to which should put the bricks in the wall, it was all settled now, and the little helpmate flew off singing out such a happy good-by it made one feel like writing a sermon on the moral effect of renunciation. After that I was sure the little helper fed his (?) mate on the nest, again singing out good-by as he flitted away. Once when he (?) brought material he found her (?) busy with what she had, and so went to the other end of the branch, and waited till she was ready for it, when he flew back and gave it to her.

It was a real delight to watch the little blue-grays at their work. Once as one of them started to fly away — I am sure this was she — she suddenly stopped to look back at the nest as if to think what she wanted to get next; or, perhaps, just to get the effect of her work at a distance, as an artist walks away from his painting; or as any mother bird would stop to admire the pretty nest that was to hold her little brood. Another time one of the gnats, — I was sure this was he, — having driven off an enemy, flipped his tail by the nest with a paternal air of satisfaction. The birds made one especially pretty picture; the little pair stood facing each other close to the nest, and the sun, filtering through the green leaves over their heads, touched them gently as they lingered near their home.

One morning when a gnat was in the nest a leaf blew down past it, startling it so it hopped out in such a hurry that the first I knew it was seated beneath the nest, flashing its tail.

Back and forth the dainty pair flew across the space of blue sky between the oak and the brush. They went so fast and carried so little it seemed as if they might have made their heads save their heels — they brought so little I could n't see that they brought anything; but I feel delicate about telling what I know about nest-making, and it may be that this was just the secret of the wonderfully compact solid walls of the nest; a little at a time, and that drilled in to stay.

When one of the small builders flew down near me — within two yards — for material, I felt greatly pleased and flattered. Her mate warned her, but she paid no particular attention to him, and with jaunty twists and turns hopped about on the dead limbs, giving hurried jabs at the cobwebs she was gathering. Once she rubbed her little cheek against a twig as if a thread of the cobweb had gotten in her eye. She dashed in among the dead leaves after something, but flew back with a start as if she had seen a ghost. She was not to be daunted, however, and after whipping her tail and peering in for a moment, hopped bravely down again. Sometimes, when collecting cobweb, the gnat would whip its tail and snap its bill snip, snip, snip, as if cutting the web with a pair of scissors.

I was amused one day by seeing a gnat fly down from the oak to the brush with what looked like a long brown caterpillar. The worm dangling from the tip of his beak was almost as large as the bird, and the little fellow had to crook his tail to keep from being overbalanced and going on his bill to the ground.

As the nest went up, the leaves hid it; but I could still see the small wings and tails flip up in the air over the edge of the cup and jerk about as the bird moulded. I watched the workers so long that I felt quite competent to build a nest myself, till happening to remember that it required gnatcatcher tools.

Ornithologists are discouraging people to wait for, and Mountain Billy got so restless under the gnat tree that he had to invent a new fly-brush for himself. On one side of the oak the branches hung low to the ground, and he pushed into the tangle till the green boughs rested on his back and he was almost hidden from view. Meanwhile I sat close beside the chaparral wall, where all sorts of sounds were to be heard, suggestive of the industries of the population hidden within the brush at my back. Hearing small footsteps, I peered in through the brown twigs, and to my delight saw a pair of stately quail walking over the ground, promenading through the brush avenues. Afterwards I caught sight of a gray animal, probably a wood rat, running down a branch behind me, and heard queer muffled sounds of gnawing.

Suddenly, looking back, I was startled to see a big ringed brown and yellow snake lying like a rope at the foot of the gnat's tree, just where I had sat. He was about four feet long, and had twenty-three rings. He started to wind into the crotch of the oak as if meaning to climb the tree, but instead, crept to a stump and festooned himself about it worming around the holes as he might do if looking for nest holes. Imagine how a mother bird would feel to have him come stealing upon her little brood in that horrid way! When he crawled over the dead leaves I noted with a shiver that he made no sound. Thinking of the gnats, I watched his every movement till he had left the premises and wormed his way off through the brush. Though quite engrossed with the gnats, it was finally forced upon me that there is more than one family in the world. The blue-gray's oak was a favored one. A pair of hang-birds had built there before the gnats came, and now two more families had come, making four for the big oak.

When first suspecting a house on the north side of the tree, I moved my chair over there. Presently a vireo with disordered breast feathers flew down on a dead twig close to the ground and leaned over with a tired anxious look, and craning her neck, turned her head on one side, and bent her eyes on the ground scrutinizingly. Then she hopped down, picked up something, threw it away, picked up another piece and flew back to her perch with it, as if to make up her mind if she really wanted that. Then her mate came, raised his crown and looked down at the bit of material with a puzzled air as if wishing he knew what to say; as if he felt he ought to be able to help her decide. But he seemed helpless and could only follow her around when she was at work, singing to her betimes, and keeping off friends or enemies who came too near. When the young hatched I noticed a still more marked difference between the nervous manners of the gnats, and the repose of vireos. While the gnat flipped about distractedly, the vireo sat calmly beside her nest, an exquisite white basket hanging under the leaves in the sun, or walked carefully over the branches looking for food for the young. Some days before finding out the facts, I suspected that the wood pewee perching on the old tree had more important business there, for the way he and his mate flew back and forth to the oak top was very pointed. So again I moved my chair. To my delight the wood pewee flew up in the tree, sat down on a horizontal crotch, and went through the motions of moulding.

There were two birds, however, that simply used the tree as a resting-place, as far as I ever knew. A hummingbird perched on the tip of a twig, looking from below like a good sized bumblebee as he preened his feathers and looked off upon the world below. At the other side of the oak a pretty pink dove perched on a sunny branch that arched against the blue sky. It sat close to the branch beside the green leaves and dressed its feathers or dozed quietly in the sun. We had other visitors that the house owners did not accept so willingly. The gnatcatchers up the sand ditch whose nest had been broken up by the thief-in-the-night did not object to brown chippies, but perhaps, if this were the same pair, they had been made suspicious by their trouble. In any case, when a brown chippie lit on a limb near the nest, quite accidentally I believe, and turned to look at the pretty structure, quite innocently I feel sure, the little gnats fell on him tooth and nail, and when he hid under the leaves where they could not reach him they fluttered above the leaves, and the moment he ventured from under cover were both at him again so violently that at the first opportunity he took to his wings. There was one curious thing about this attack and expulsion; the gnats did not utter a word during the whole affair! I had never known them to be silent before when anything was going on — rarely when there was n't.

Another morning when I rode in there was a great commotion up in the oak. A chorus of small scolding voices, and a fluttering of little wings among the branches told that something was wrong, while a large form moving deliberately about in the tree showed the intruder to be a blue jay! Aha! the gossips would wag their heads. I disapprove of gossip, but as a truthful reporter am obliged to say that I saw the blue jay pitch down into the brush with something white in his bill — perhaps a cocoon — and that thereupon a great weeping and wailing arose from the little folk up in the treetop. A big brown California chewink stood by and watched the — robbery (?), great big fellow that he was; and not once offered to take the little fellows' part. I felt indignant. Why did n't he pitch into the big bully and drive him off before he had stolen the little birds' egg — if it was an egg. A grosbeak called *ick'* from the treetop, but thought he 'd better not meddle; and — it was a pair of wren-tits who looked out from a brush screen and then skulked off, chuckling to themselves, I dare say, that some one else was up to their tricks. It gave my faith in birds a great shock, this, together with the pillage of the gnat's nest by the thief-in-the-night. My spleen was especially turned against the brown chewink; he certainly was a good fighter, and might at least have helped to clear the neighborhood of such a suspicious character.

Where did the egg — if it was an egg — come from? The vireos and pewees and gnats were still building, I reflected thankfully, though trembling for their future; and fortunately the hangbird had young. Perhaps the jay had found a nest that I could not discover.

After that, things went on quietly for several days. The gnats got through with their building, and went off for a holiday until it should be time to begin brooding. They flitted about the branches warbling, as if having nothing special to do; dear little souls, at work as at play, always together. One of them unexpectedly found himself near me one day; but when he saw it was only I, whipped his tail and exclaimed "*Oh, it's you. I'm not afraid.*"

This peace and quietness, however, did not last. The gnats' house was evidently haunted, and they did not like — blue — ghosts. One morning when

I got to the oak it was all in a hubbub, and the vireo was scolding loudly at a blue jay. When the giant pitched into the brush the wren-tit chattered, and I thought perhaps the jay was teaching him how it feels to have a shoe pinch. A few moments later I was amazed to see a gnat jab at the wall till it got a bill full of material and then fly off to the brush with it! My little birds had moved! Evidently the neighborhood was too exciting for them. More than ten days of hard work — no one can tell how hard until after watching a gnatcatcher build — had been spent in vain on this nest; and if, as suspected, this was their second, how much more work did that mean? It was a marvel that the birds could get courage to start in again, especially if they had had two homes broken up already.

From my position at the big oak I could see that the gnats were carrying the frame of the old house to a small oak in the brush. The wood pewee had moved too, and to my surprise and pleasure I found it had begun its nest on a branch under the gnats, so that both families could be watched at the same time. I nearly got brushed off the saddle promenading through the stiff chaparral to find a place where the nests could be seen from the ground; but when at last successful, I too, like the rest of the old oak's floating population, moved to pastures new. Hanging my chair on the saddle, I made Billy carry it for me; then I buckled the reins around the trunk of the oak and withdrew into the brush to watch my birds. It was a cozy little nook, from which Billy could be heard stamping his feet to shake off the flies. The little crack in the chaparral was a pleasant place to sit in, protected as it was from the wind, with the sun only coming in enough to touch up the brown leaves on the ground and warm the fragrant sage, bringing out its delicious spicy aromatic smell.

The pewee did not altogether relish having us established under its vine and fig-tree. When it saw Billy under the tree it whistled, and the bit of grass it had brought for its nest went sailing down to the brush disregarded. It did not think us as bad as the blue jay, however, for it came back with a long stem of grass in its bill, and, lighting on a high branch, called *pee-ree*. To be sure, when it had gone to the nest and I was inconsiderate enough to turn a page in my note-book, it dashed off. But if murder will out, so will good intentions; and before long the timid bird was brooding its nest with Billy and me for spectators.

The gnat's nest here was so much lower than the other one that it was much easier to watch. The first day the birds built rapidly. One of them got his spider's web from beside the pewee's nest, when the pewee was away. He started to go for it once after the owner had returned, caught sight of him, stopped short, and much to my amusement concluded to sit down and preen his feathers! The pewee had one special bare twig of his own that he used for a perch, and when the gnat seated himself there in his neighbor's absence he looked so small that I realized what a mite of a bird he really was. He sometimes sat there and talked while his mate moulded the nest.

When the gnats got to brooding, many of the same pretty performances were repeated that had marked the first nest of all, up in the sand ditch. When the bird on the nest hopped out and called, "Come, come," its mate, who had been wandering around in the sunny green treetop, called out in sweet tones, "Good-by, good-by."

When waiting for the gnats to do something, I heard a little sound in the oak brush by my side, and, looking through the brown branches, saw a wren-tit come hopping toward me. It came up within three feet of me, near enough to see its bright yellow eyes. I began to wonder if it had a nest near by, and felt my prejudices melting away and my heart growing tender. Some thieves are very honest fellows; it is largely a difference in ethical standards! I began to feel a keen interest in the bird and its affairs, for the wren-tit was really a most original bird, and one I was especially anxious to study.

My newly awakened interest was not chilled by any second tragedy; all went well with the little blue-grays. The day the gnat's eggs hatched, the old folks performed most ludicrously. Perhaps they were young parents, and this being their first brood, maternal and paternal love had not yet blinded their eyes to the ridiculous; so that they looked down on these skinny, squirming, big-eyeballed prodigies with mingled emotions. It looked very much as if they were surprised to find that their smooth pretty eggs had suddenly turned into these ugly, weak, hungry things they did not know what to do with. At first it seemed that something must be wrong at the nest; the little gnat shook her wings and tail beside it as if afraid of soiling herself; and when she hopped into it, jerked out again and flitted around distractedly. Every time the birds looked into the nest they got so excited that, had they been girls, they surely would have hopped up and down wringing their hands. I laughed right out alone in the brush, they acted so absurdly.

They began feeding the nestlings in the most remarkable way I had ever witnessed. When the young mother was on the nest her mate came and brought her the food, whereupon, instead of jumping off the nest and feeding the young in the conventional way, she simply raised up on her feet and, apparently, poked the food backwards into the bills of the young under her breast! Even when the gnats got to feeding more in the ordinary way, they did it nervously. They fed as if expecting the young to bite them. They would fly up on the branch beside the nest, give a jab down at the youngsters, whip tails and flee. You would have thought the young parents had been playing house before, and their dolls had suddenly turned into live hungry nestlings.

I watched this family till the house was deserted, and I had to ride along a line of brush before finding them. The young were now pretty silvery-breasted creatures who sat up in a small oak while the old birds hunted through the brush for food for them. Though I rode Billy into the chaparral after them, and got near enough to see the black line over the bill of the father bird, they did not mind, but hunted away quite unconcernedly; for we had been through many things together, and were now old and fast friends.

V.

LITTLE PRISONERS IN THE TOWER.

I HAD not spent many days in The Little Lover's dooryard before realizing that there was something in the wind. If an inoffensive person fancies sitting in the shade of a sycamore with her horse grazing quietly beside her, who should say her nay? If, at her approach, a — feathered — person steals away to the top of the highest, most distant oak within sight and, silent and motionless, keeps his eye on her till she departs; if, as she innocently glances up at the trees, she discovers a second — feathered — person's head extended cautiously from behind a trunk, its eyes fixed on hers; or if, as she passes along a — sycamore — street, a person comes to a window and cranes his neck to look at her, and instantly leaves the premises; then surely, as the world wags, she is quite justified in having a mind of her own in the matter. Still more, when it comes to finding chips under a window — who could do aught but infer that a carpenter lived within? Not I. And so it came about that I discovered that one of the apartments in the back of the wren sycamore

California Woodpecker.
(One half natural size.)

had been rented by a pair of well-meaning but suspicious California woodpeckers, first cousins of the eastern red-heads.

It is unpleasant to be treated as if you needed detectives on your track. It strains your faith in human nature; the rest of the world must be very wicked if people suspect such extremely good creatures as you are! And then it reflects on the detectives; it shows them so lacking in discernment. Nevertheless, "A friend should bear his friend's infirmities," and I was determined to be friends with the woodpeckers. One of them kept me waiting an hour one morning. When I first saw it, it was on its tree trunk, but when it first saw me, it promptly left for parts unknown. I stopped at a respectful distance from its tree — several rods away — and threw myself down on the warm sand in the

Red-headed Woodpecker — Eastern.
(One half natural size.)

bed of the dry stream, between high hedges of exquisite lemon-colored mustard. Patient waiting is no loss, observers must remember if they would be consoled for their lost hours. In this case I waited till I felt like a lotus-eater who could have stayed on forever. A dove brooded her eggs on a branch of the spreading sycamore whose arms were outstretched protectingly above me; the sun rested full on its broad leaves, and bees droned around the fragrant mustard, whose exquisite golden flowers waved gently against a background of soft blue California sky.

But that was not the last day I had to wait. It was over a month before the birds put any trust in me. The nest hole was excavated before the middle of May; on June 15 I wrote in my notebook, "The woodpecker has gotten so that when I go by she puts her head out of the window, and when I speak to her does not fly away, but cocks her head and looks down at me."[1] That same morning the bird actually entered the nest in my presence. She came back to her sycamore while I was watching the wrens, and flew right up to the mouth of the nest. She was a little nervous. She poked in her bill, drew it back; put in her head, drew that back; then swung her body partly in; but finally the tip of her tail disappeared down the hole.

[1] The difference in the dress of the woodpeckers is so slight that the sexes were not distinguished at this nest.

The next morning, in riding by, I heard weak voices from the woodpecker mansion. If young were to be fed, I must be on hand. Such luxurious observing! Riding Mountain Billy out into the meadow, I dismounted, and settled myself comfortably against a haycock with the bridle over my arm. It was a beautiful quiet morning. The night fog had melted back and the mountains stood out in relief against a sky of pure deep blue. The line of sycamores opposite us were green and still against the blue; the morning sun lighting their white trunks and framework. The songs of birds filled the air, and the straw-colored field dotted with haycocks lay sunning under the quiet sky. In the East we are accustomed to speak of "the peace of evening," but in southern California in spring there is a peculiar interval of warmth and rest, a langorous pause in the growth of the morning, between the disappearance of the night fog and the coming of the cool trade wind, when the southern sun shines full into the little valleys and the peace of the morning is so deep and serene that the labor of the day seems done. Nature appears to be slumbering. She is aroused slowly and gently by the soft breaths that come in from the Pacific. On this day I watched the awakening. Up to this time not a grass blade had stirred, but while I dreamed a brown leaf went whirling to the ground, the stray stalks of oats left from the mowing began to nod, and the sycamore branches commenced to sway. Then the breeze swelled stronger, coming cool and fresh from the ocean; the yellow primroses, around which the hummingbirds whirred, bowed on their stately stalks, and I could hear the wind in the moving treetops.

Mountain Billy grazed near me till it occurred to him that stubble was unsatisfactory, when he betook him to my haycock. Though I lectured him upon the rights of property and enforced my sermon with the point of the parasol, he was soon back again, with the amused look of a naughty boy who cannot believe in the severity of his monitor; and later, I regret to state, when I was engrossed with the woodpeckers, a sound of munching arose from behind my back.

The woodpeckers talked and acted very much like their cousins, the red-heads of the East. When they went to the nest they called *chuck′-ah* as if to wake the young, flying away with the familiar rattling *kit-er′r′r′r′*. They flew nearly half a mile to their regular feeding ground, and did not come to the nest as often as the wrens when bringing up their brood. Perhaps they got more at a time, filling their crops and feeding by regurgitation, as I have seen waxwings do when having a long distance to go for food.

I first heard the voices of the young on June 16; nearly three weeks later, July 6, the birds were still in the nest. On that morning, when I went out to mount Billy, I was shocked to find the body of one of the old woodpeckers on the saddle. I thought it had been shot, but found it had been picked up in the prune orchard. That afternoon its mate was brought in from the same place. Probably both birds had eaten poisoned raisins left out for the gophers. The dead birds were thrown out under the orange-trees near the house, and not many hours afterward, when I looked out of the window, two turkey vultures were sitting on the ground, one of them with a pathetic little black wing in his bill. The great black birds seemed horrible to me, — ugly, revolting creatures. I went outside to see what they would do, and after craning their long red necks at me and stalking around nervously a few moments they flew off.

Now what would become of the small birds imprisoned in the tree trunk, with no one to bring them food, no one to show them how to get out, or, if they were out, to feed them till they had learned how to care for themselves? Sad and anxious, I rode down to the sycamore. I rapped on its trunk, calling *chuck'-ah* as much like the old birds as possible. There was an instant answer from a strong rattling voice and a weak piping one. The weak voice frightened me. If that little bird's life were to be saved, it was time to be about it. The ranchman's son was pruning the vineyard, and I rode over to get him to come and see how we could rescue the little prisoners.

On our way to the tree we came on a gopher snake four feet long. It was so near the color of the soil that I would have passed it by, but the boy discovered it. The creature lay so still he thought it was dead; but as we stood looking, it puffed itself up with a big breath, darted out its tongue, and began to move off. I watched to see how it made the straight track we so often saw in the dust of the roads. It bent its neck into a scallop for a purchase, while its tapering tail made an S, to furnish slack; and then it pulled the main length of its body along straight. It crawled noiselessly right to the foot of the woodpecker tree, but was only hunting for a hole to hide in. It got part way down one hole, found that it was too small, and had to come backing out again. It followed the sand bed, taking my regular beat, from tree to tree! To be sure, gopher snakes are harmless, but they are suggestive, and you would rather their ways were not your ways.

Although the little prisoners welcomed us as rescuers should be welcomed, they did it by mistake. They thought we were their parents. At the first blow of the axe their voices hushed, and not a sound came from them again. It seemed as if we never should get the birds out.

It looked easy enough, but it was n't. The nest was about twelve feet above the ground. The sycamore was so big the boy could not reach around it, and so smooth and slippery he could not get up it, though he had always been a good climber. He clambered up a drooping branch on the back of the tree, — the nest was in front, — but could not swing himself around when he got up. Then he tried the hollow burned at the foot of the tree. The charred wood crumbled beneath his feet, but at last, by stretching up and clinging to a knothole, he managed to reach the nest.

As his fingers went down the hole, the young birds grabbed them, probably mistaking them for their parents' bills. "Their throats seem hot," the boy exclaimed; "poor hungry little things!" His fingers would go through the nest hole, but not his knuckles, and the knothole where he steadied himself was too slippery to stand on while he enlarged the hole. It was getting late, and as he had his chores to do before dark I suggested that we feed the birds and leave them in the tree till morning; but the rescuer exclaimed resolutely, "We'll get them out to-night!" and hurried off to the ranch-house for a step-ladder and axe.

The ladder did not reach up to the first knothole, four or five feet below the nest; but the boy cut a notch in the top of the knot and stood in it, practically on one foot, and held on to a small branch with his right hand — the first limb he trusted to broke off as he caught it — while with the left hand he hacked away at the nest hole. It was a ticklish position and genuine work, for the wood was hard and the hatchet dull.

I stood below holding the carving-knife, — we had n't many tools on the ranch, — and as the boy worked he entertained me with an account of an accident that happened years before, when his brother had chopped off a branch and the axe head had glanced off, striking the head of the boy who was watching below. I stood from under as he finished his story, and inquired with interest if he were sure his axe head was tight! Before the lad had made much impression on the hard sycamore, he got so tired and looked so white around the mouth that I insisted on his getting down to rest, and tried to divert him by calling his attention to the sunset and the voices of the quail calling from the vineyard. When he went up again I handed him the carving-knife to slice off the thinner wood on the edge of the nest hole, warning him not to cut off the heads of the young birds.

At last the hole was big enough, and, sticking the hatchet and knife into the bark, the lad threw one arm around the trunk to hold on while he thrust his hand down into the nest. "My, what a deep hole!" he exclaimed. "I don't know as I can reach them now. They 've gone to the bottom, they 're so afraid." Nearly a foot down he had to squeeze, but at last got hold of one bird and brought it out. "Drop him down," I cried, "I 'll catch him," and held up my hands. The little bird came fluttering through the air. The second bird clung frightened to the boy's coat, but he loosened its claws and dropped it down to me. What would the poor old mother woodpecker have thought had she seen these first flights of her nestlings!

I hurried the little scared brothers under my jacket, my best substitute for a hollow tree, and called *chuck'-ah* to them in the most woodpecker-like tones I could muster. Then the boy shouldered the ladder, and I took the carving-knife, and we trudged home triumphant; we had rescued the little prisoners from the tower!

When we had taken them into the house the woodpeckers called out, and the cats looked up so savagely that I asked the boy to take the birds home to his sister to keep till they were able to care for themselves. On examining them I understood what the difference in their voices had meant. One of them poked his head out of the opening in my jacket where he was riding, while the other kept hidden away in the dark; and when they were put into my cap for the boy to carry home, the one with the weak voice disclosed a whitish bill — a bad sign with a bird — and its feeble head bent under it so weakly that I was afraid it would die.

Three days later, when I went up to the lad's house, it was to be greeted by loud cries from the little birds. Though they were in a box with a towel over it, they heard all that was going on. Their voices were as sharp as their ears, and they screamed at me so imperatively that I hurried out to the kitchen and rummaged through the cupboards till I found some food for them. They opened their bills and gulped it down as if starving, although their guardian told me afterwards that she had fed them two or three hours before.

When held up where the air could blow on them, they grew excited; and one of them flew down to the floor and hid away in a dark closet, sitting there as contentedly as if it reminded him of his tree trunk home.

I took the two brothers out into the sitting-room and kept them on my lap for some time, watching their interesting ways. The weak one I dubbed Jacob, which is the name the people of the valley had given the woodpeckers from the sound of their cries; the stronger bird I called Bairdi, as 'short' for *Melanerpes formicivorous bairdi* — the name the ornithologists had given them.

Jacob and Bairdi each had ways of his own. When offered a palm, Bairdi, who was quite like 'folks,' was content to sit in it; but Jacob hung with his claws clasping a little finger as a true woodpecker should; he took the same pose when he sat for his picture. Bairdi often perched in my hand, with his bill pointing to the ceiling, probably from his old habit of looking up at the door of his nest. Sometimes when Bairdi sat in my hand, Jacob would swing himself up from my little finger, coming bill to bill with his brother, when the small bird would open his mouth as he used to for his mother to feed him. Poor little orphans, they could not get used to their changed conditions!

They did other droll things just as their fathers had done before them. They used to screw their heads around owl fashion, a very convenient thing for wild birds who cling to tree trunks and yet need to know what is going on behind their backs. Once, on hearing a sudden noise, one of them ducked low and drew his head in between his shoulders in such a comical way we all laughed at him.

I often went up to the ranch to visit them. We would take them out under a big spreading oak beside the house, where the little girl's mother sat with her sewing, and then watch the birds as we talked. When we put them on the tree trunk, at first they did not know what to do, but soon they scrambled up on the branches so fast their guardian had to climb up after them for fear they would get away. Poor little Jacob climbed as if afraid of falling off, taking short hops up the side of the tree, bending his stiff tail at a sharp angle under him to brace himself against the bark. Bairdi, his strong brother, was less nervous, and found courage to catch ants on the bark. Jacob did a pretty thing one day. When put on the oak, he crept into a crack of the bark and lay there fluffed up against its sides with the sun slanting across, lighting up his pretty red cap. He looked so contented and happy it was a pleasure to watch him. Another time he started to climb up on top of my head and, I dare say, was surprised and disappointed when what he had taken for a tree trunk came to an untimely end. When we put the brothers on the grass, one of them went over the ground with long hops, while the other hid under the rocking-chair. One bird seemed possessed to sit on the white apron worn by the little girl's mother, flying over to it from my lap, again and again.

The woodpeckers had brought from the nest a liking for dark, protected places. Bairdi twice clambered up my hair and hung close under the brim of my black straw hat. Another time he climbed up my dress to my black tie and, fastening his claws in the silk, clung with his head in the dark folds as if he liked the shade. I covered the pretty pet with my hand and he seemed to enjoy it. When I first looked down at him his eyes were open, though he kept very still; but soon his head dropped on my breast and he went fast asleep, and would have had a good nap if Jacob had not called and waked him up.

Jacob improved so much after the first few days — and some doses of red pepper — that we had to look twice to tell him from his sturdy brother. He certainly ate enough to make him grow.

The birds liked best to be fed with a spoon; probably it seemed more like a bill. After a little, they learned to peck at their food, a sign I hailed eagerly as indicative of future self-support; for with appetites of day laborers and no one to supply their wants, they would have suffered sorely, poor little orphans! Sometimes, when they had satisfied their first hunger, they would shake the bread from their bills as if they didn't like it and wanted food they were used to.

When one got hungry he would call out, and then his brother would begin to shout. The little tots gave a crooning gentle note when caressed, and a soft cry when they snuggled down in our hands or cuddled up to us as they had done under their mother's wing. Their call for food was a sibilant chirr, and they gave it much oftener than any of the grown-up woodpecker notes. But they also said *chuck'-ah* and rattled like the old birds.

I was glad there were two of them so they would not be so lonely. If separated they showed their interest in each other. If Bairdi called, Jacob would keep still and listen attentively, raising his topknot till every microscopic red feather stood up like a bristle, when he would answer Bairdi in a loud manly voice.

It was amusing to see the small birds try to plume themselves. Sometimes they would take a sudden start to make their toilettes, and both work

JACOB AND BAIRDI VISITING THE OLD NEST TREE

away vigorously upon their plumes. It was comical to see them try to find their oil glands. Had the old birds taught them how to oil their feathers while they were still in the nest? They were thickly feathered, but when they reached back to their tails the pink skin showed between their spines and shoulders, giving a good idea of the way birds' feathers grow only in tracts.

When the little princes were about a month old, I arranged with a neighboring photographer to have them sit for their picture. He drove over to the sycamore, and the lad who had rescued the prisoners took them down to keep their appointment. One of them tried to tuck its head up the boy's sleeve, being attracted by dark holes. While we were waiting for the photographer, the boy put Jacob in a hollow of the tree, where he began pecking as if he liked it. He worked away till he squeezed himself into a small pocket, and then, with his feathers ruffled up, sat there, the picture of content. Indeed, the little fellow looked more at home than I had ever seen him anywhere. The rescuer was itching to put the little princes back in their hole, to see what they would do, but I would n't listen to it, being thankful to have gotten them out once.

When Bairdi was on the bark and Jacob was put below him, he turned his head, raised his red cap, and looked down at his brother in a very winning way.

Soon the photographer came, and asked, "Are these the little chaps that try to swallow your fingers?" We were afraid they would not sit still enough to get good likenesses, but we had taken the precaution to give them a hearty breakfast just before starting, and they were too sleepy to move much. In the picture, Jacob is clinging to the boy's hand in his favorite way, and Bairdi is on the tree trunk.

Mountain Billy pricked up his ears when he discovered the woodpeckers down at the sycamore, but he often saw them up at the ranch and took me to make a farewell call on them before I left for the East. We found the birds perched on the tobacco-tree in front of the ranch-house, with a tall step-ladder beside it so the little girl could take them in at night. Their cup of bread and milk stood on the ladder, and when I called them they came over to be fed. They were both so strong and well that they would soon be able to care for themselves, as their fathers had done before them. And when they were ready to fly, they might have help; for an old woodpecker of their family — possibly an unknown uncle — had been seen watching them from the top of a neighboring oak, and may have been just waiting to adopt the little orphans. In any case, however they were to start out in the world, it was a great satisfaction to have rescued them from their prison tower.

VI.

HINTS BY THE WAY.

On our way back and forth along the line of oaks and sycamores belonging to the little prisoners, the little lover, and the gnatcatchers, Mountain Billy and I got a good many hints, he of places to graze, and I of new nests to watch.

While waiting for the woodpeckers one day I saw a small brownish bird flying busily back and forth to some green weeds. She was joined by her mate, a handsome blue lazuli bunting, even more beautiful than our lovely indigo bunting, and he flew beside her full of life and joy. He lit on the side of a cockle stem, and on the instant caught sight of me. Alas! he seemed suddenly turned to stone. He held onto that stalk as if his little legs had been bars of iron and I a devouring monster. When he had collected his wits enough to fly off, instead of the careless gay flight with which he had come out through the open air, he timidly kept low within the cockle field, making a circuitous way through the high stalks.

He could be afraid of me if he liked, I thought, — for after a certain amount of suspicion an innocent person gets resentful; at any rate, I was going to see that nest. Creeping up cautiously when the mother bird was away, so as not to scare her, and carefully parting the mallows, I looked in. Yes, there it was, a beautiful little sage-green nest of old grass laid in a coil. I felt as pleased as if having a right to share the family happiness.

After that I watched the small worker gather material with new interest, knowing where she was going to put it. She worked fast, but did not take the first thing she found, by any means. With a flit of the wing she went in nervous haste from cockle to cockle, looking eagerly about her. Jumping down to the ground, she picked up a bit of grass, threw it down dissatisfied, and turned away like a person looking for something. At last she lit on the side of a thistle, and tweaking out a fibre flew with it to the nest.

When the house was done, one morning in passing I leaned down from the saddle, and through the weeds saw her brown wings as she sat on the nest. A month after the first encounter with the father lazuli, I found him looking at me around the corner of a cockle stalk, and in passing back again caught him singing full tilt, though his bill was full of insects! After we had turned our backs, I looked over my shoulder and had the satisfaction of seeing him take his beakful to the nest. You could n't help admiring him, for though not a warrior who would snap his bill over the head of an enemy of his home, he had a gallant holiday air with his blue coat and merry song, and you felt sure his little brown mate would get cheer and courage enough from his presence to make family dangers appear less frightful. Even this casual acquaintance with the little pair gave me a new and tender interest in all of their name I might know in future.

While watching the lazulis from the sycamores, on looking up on a level with Billy's ears, I discovered a snug canopied nest held by a jointed branch of the twisted tree, as in the palm of your hand. It was as if the old sycamore were protecting the little brood, holding it secure from all dangers. Looking at the nest, I spied a brown tail resting against the limb, and then a small brown head was raised to look at me from between the leaves. It was the little bird whose sweet home-like song had so cheered my heart in this far-away land, the home song sparrow, dearer than all the birds of California. It was such a pleasure to find her that I sat in the saddle and talked to the pretty bird while she brooded her eggs under the green leaves.

The next time we went down to the sycamore the bird was away, and it seemed as if the tree had been deserted. It was empty and uninteresting. Again I came, and this time the father song sparrow sang blithely in the old tree, while his gentle mate went about looking for food for her brood. Her little birds had come! How happy and full of business she seemed! She ran nimbly over the ground, weaving in and out between the stalks of the oats and the yellow mustard, as if there were paths in her forest. When she had to run across the sand bed, out in open sight, she put up her tail, held her wings tight at her sides, and scudded across. Then with the sunlight through the leaves dappling her back, she ran around the foot of the sycamore. She had something in her bill, and with a happy chirp was off to her brood.

There was another family abroad on our beat. When riding past the little lover's, I heard voices of young birds beyond, and rode out to the oak in the middle of the field from which they came, to see who it was. It was a surprise to find a family of full-fledged blue jays — a surprise, because the jays had been terrorizing the small birds of the neighborhood till it seemed strange to think they had any family life themselves. I had come to feel that they were great hobgoblins going about seeking whom they could devour; but such harsh judgments are usually false, whether of birds or beasts, and I was convinced against my will on hearing the tender tone in which the old jays called to their young.

To be sure, they were imperative in their commands. As I rode around the tree, one of them looked at me sharply and proceeded to take measures to protect his brood. When one of the children told me where he was, his parent promptly flew over and shouted in his ear, "Be quiet!" with such a ring of command that an unbroken hush followed. Moreover, when one child, probably a greedy one, teased for food, its parent ran down the branch to drive it off; and in some way best known to themselves the old birds hushed up the boisterous young ones and spirited them out of my sight. But all these things were in line with good family government and the best interests of the children, and were more than atoned for by the soft gentle notes the old birds used when they were leading around their cherished brood out of harm's way.

VII.

AROUND OUR RANCH-HOUSE.

CLOSE up under the hills, the old vine-covered ranch-house stood within a circle of great spreading live oaks. The trees were full of noisy, active blackbirds — Brewer's blackbirds, relatives of the rusty that we know in New York. The ranchman told me that they always came up the valley from the vineyard to begin gathering straws for their nests on his brother's birthday, the twenty-fifth of March. After that time it was well for passers below to beware. If an unwary cat, or even a hen or turkey gobbler, chanced under the blackbirds' tree, half a dozen birds would dive down at it, screaming and scolding till the intruders beat an humble retreat. But the blackbirds were not always the aggressors. I heard a great outcry from them one day, and ran out to find them collecting at the tree in front of the house. A moment later a hawk flew off with a young nestling, and was followed by an angry black mob.

One pair of the blackbirds nested in the oak by the side of the house, over the hammock. Though making themselves so perfectly at home on the premises, driving off the ranchman's cats and gobblers, and drinking from his watering-trough, if they were taken at close quarters, with young in their nests, the noisy birds were astonishingly timid. One could hardly understand it in them.

One afternoon I sat down under the tree to watch them. Mountain Billy rested his bridle on my knee, and the ranchman's dog came out to join us; but the mother blackbird, though she came with food in her bill and started to walk down the branch over our heads, stopped short of the nest when her eye fell on us. She shook her tail and called *chack*, and her mate, who sat near, opened wide his bill and whistled *chee*. The small birds were hungry and grew impatient, seeing no cause for delay, so raised their three fuzzy heads above the edge of the nest and sent imperative calls out of their three empty throats. As the parents did not answer the summons, the young dozed off again, but when the old ones did get courage to light near the nest there was such a rousing chorus that they flew off alarmed for the safety of their clamorous brood. After that outbreak, it seemed as if the mother bird would never go back to her children; but finally she came to the tree and, after edging along falteringly, lit on a branch above them. The instant she touched foot, however, she was seized with nervous qualms and turned round and round, spreading her tail fan-fashion, as if distracted.

To my surprise, it was the father bird who first went to the nest, though he had the wit to go to it from the outside of the tree, where he was less exposed to my dangerous glance. I wondered whether it was mother love that kept her from the nest when he ventured, or merely a case of masculine common-sense versus nerves. How birds could imagine more harm would be done by going to the nest than by making such a fuss five feet away from it was a poser to me. Perhaps they attribute the same intelligence to us that some of us do to them!

While the blackbirds were making such a time over our heads, I watched the hummingbirds buzzing around the petunias and pink roses under the ranch-house windows, and darting off to flutter about the tubular flowers of the tobacco-tree by the well. One day the small boy of the family climbed up to the hummingbird's nest in the oak "to see if there were eggs yet," and the frightened brood popped out before his eyes. His sister caught one of them and brought it into the house. When she held it up by the open door the tiny creature spread its little wings and flew out into the vines over the window. The child was so afraid its mother would not find it she carried it back to its oak and watched till the mother came with food. The hummers were about the flowers in front of the windows so much that when the front door was left open they often came into the room.

In an oak behind the barn I found a humming-bird's nest, and, yielding to temptation, took out the eggs to look at them. In putting them back one slipped and dropped on the hard ground, cracking the delicate pink shell as it fell. The egg was nearly ready to hatch, and I felt as guilty as if having killed a hummingbird.

When in the hammock under the oak one day, I saw a pair of the odd-looking Arizona hooded orioles busily going and coming to a drooping branch on the edge of the tree. They had a great deal to talk about as they went and came, and when they had gone I found, to my great satisfaction, that they had begun a nest. They often use the gray Spanish moss, but here had found a good substitute in the orange-colored parasitic vine of the meadows known among the people of the val-

Arizona Hooded Oriole.
(One half natural size.)

Baltimore Oriole — Eastern.
(One half natural size.)

ley as the 'love-vine' (dodder). The whole pocket was composed of it, making a very gaudy nest.

Linnets nested in the same old tree. Indeed, it is hard to say where these pretty rosy house finches, cousins of our purple finches, would not take it into their heads to build. They nested over the front door, in the vines over the windows, in the oaks and about the outbuildings, and their happy musical songs rang around the ranch-house from morning till night. As I listened to their merry roundelay day after day during that beautiful California spring, it sounded to me as though they said, "*How-pretty-it-is'-out, how-pretty-it-is'-out, how-pretty-it-is' !*" The linnets are ardent little wooers, singing and dancing before the indifferent birds they would win for their mates. I once saw a rosy lover throw back his pretty head and hop about before his brown lady till she was out of patience and turned her back on him. When that had no effect, she opened her bill, spread her wings, and leaned toward him as if saying, "If you don't stop your nonsense, I'll —" But the fond linnets' gallantry and tenderness are not all spent in the wooing. When the mother bird was brooding her nest over our front door, her crimson-throated mate stood on the peak of the ridgepole above and sang blithely to her, turning his head and looking down every little while to make sure that she was listening to his pretty prattle.

One of the birds that nested in the trees by the ranch-house was the bee-bird, who was soft gray above and delicate yellow below, instead of dark gray above and shining white below, like his eastern relative, the kingbird. The birds used to perch on the bare oak limbs, flycatching. It was interesting to watch them. They would fly obliquely into the air and then turn, with bills bristling with insects, and sail down on outstretched wings, their square tails set so that the white outer feathers showed to as good advantage as the white border of the kingbird's does in similar flights. They made a bulky untidy nest in the oaks by the barn, using a quantity of string borrowed from the ranchman. Their voices were high-keyed and shrill with an impatient emphasis, and at a distance suggested the shrill yelping of the coyote. *Kee'-ah, kee-kee' kee'-ah*, they would cry. The wolves were so often heard around the ranch-house that in the early morning I have sometimes mistaken the birds for them.

One of the favorite hunting-grounds of the bee-birds was the orchard, where they must have done a great deal of good destroying insects. They were quarrelsome birds, and were often seen falling through the air fighting vigorously. I saw one chase a sparrow hawk and press it so hard that the hawk cried out lustily. The ranchman's son told me of one bee-bird who defended his nest with his life. Two crows lit in a tree where the flycatcher had a nest containing eggs. The crows had difficulty in getting to the tree to begin with, for the bee-birds fought them off; and though they lighted, were soon dislodged and chased down the vineyard. The man was at work there, and as the procession passed over his head the bee-bird dove at the crow; the crow struck back at him, crushing his skull, and the flycatcher dropped through the air, dead! The other bee-bird followed its dead mate to the ground, and then, without a cry, flew to a tree and let the crows go on their way.

The bee-bird was one of the noisiest birds about the ranch-house, but commoner than he; in fact, the most abundant bird, next to the linnet and blackbird, was the California chewink, or, as the ranchman appropriately called him, the 'brown chippie;' for he does not look like the handsome chewink we know, but is a fat, dun brown bird with a thin *chip* that he utters on all occasions. He is about the size of the eastern robin, and, except when nesting, almost as familiar. There were brown chippies in the door-yard, brown chippies around the barns, and brown chippies in the brush till one got tired of the sight of them.

The temptations that come to conscientious observers are common to humanity, and one of the subtlest is to undervalue what is at hand and overvalue the rare or distant. Unless a bird is peculiarly interesting, it requires a definite effort to sit down and study him in your own door-yard, or where he is so common as to be an every-day matter. The chippies were always sitting around, scratching, or picking up seeds; or else quarreling among themselves. Feeling that it was my duty to watch them, I reasoned with myself, but they seemed so mortally dull and uninteresting it was hard work to give up any time to them. When they went to nesting, their wild instincts asserted themselves, and they hid away so closely I was never sure of but one of their nests, and that only by most cautious watching. Then for the first time they became interesting! To my surprise, one day I heard a brown chippie lift up his voice and sing. It was in a sunny grove of oaks, and though his song was a queer squeaky warble, it had in it a good deal of sweetness and contentment; for the bird seemed to find life very pleasant. The ranchman's son told me that up in the canyons at dusk he had sometimes heard towhee concerts, the birds answering each other from different parts of the canyon.

California Chewink.
(One half natural size.)

Eastern Chewink.
(One half natural size.)

There was a nest in the chaparral which probably belonged to these chewinks. It was in a mass of poison ivy that had climbed up on a scrub-oak. I spent the best part of a morning waiting for the birds to give in their evidence. Brown sentinels were posted on high bare brush tops, where they chipped at me, and once a brown form flew swiftly away from the nest bush; but like most people whose conversation is limited to monosyllables, the towhees are good at keeping a secret. While watching for them, I heard a noise that suggested angry cats spitting at each other; and three jack-rabbits came racing down the chaparral-covered knoll. One of them shot off at a tangent while the other two trotted along the openings in the brush as if their trails were roads in a park. Then a cottontail rabbit came out on a spot of hard yellow earth encircled by bushes, and lying down on its side kicked up its heels and rolled like a horse; after which the pretty thing stretched itself full length on the ground to rest, showing a pink light in its ears. After a while it got up, scratched one ear, and with a kick of one little furry leg ran off in the brush. Another day, when I sat waiting, I saw a jack-rabbit's ears coming through the brush. He trotted up within a few feet, when he stopped, facing me with head and ears up; a noble-looking little animal, reminding me of a deer with antlers branching back. He stood looking at me, not knowing whether to be afraid or not, and turning one ear trumpet and then the other. But though smiling at him, I was a human being, there was no getting around that; and after a few undecided hops, this way and that, he ran off and disappeared in the brush. Near where he had been was a spot where a number of rabbit runways came to a centre, and around it the rabbit council had been sitting in a circle, their footprints proved.

Brown chippies were not much commoner around the ranch-house than western house wrens were, but the big prosaic brown birds seemed much more commonplace. The wrens were strongly individual and winning wherever they were met. They nested in all sorts of odd nooks and corners about the buildings. One went so far as to take up its abode in the wire-screened refrigerator that stood outside the kitchen under an oak! Another pair stowed their nest away in an old nose-bag hanging on a peg in the wine shed; while a third lived in one of the old grape crates piled up in the raisin shed.

The crate nest was delightful to watch. The jolly little birds, with tails over their backs and wings hanging, would sing and work close beside me, only three or four feet away. They would look up at me with their frank fearless eyes and then squeeze down through their crack into the crate, and sit and scold inside it—such an amusing muffled little scold! The nest was so astonishingly large I was interested to measure it. Twigs were strewn loosely over one end of the box, covering a square nearly sixteen inches on a side. The compact high body of the nest measured eight by ten inches, and came so near the top of the crate that the birds could just creep in under the slats. Some of the twigs were ten inches long, regular broom handles in the bills of the short bobbing wrens. One of the birds once appeared with a twig as long as itself. It flew to the side of a beam with it, at sight of me, and stood there balancing the stick in its bill, in pretty fashion. Another time it flew to the peak of the shed to examine an old swallow's nest now occupied by linnets, and amused itself throwing down its neighbors' straws — the naughty little rogue!

Such jolly songsters! They were fairly bubbling over with happiness all the time. They had an old stub in front of the shed that might well have been called the singing stub, for they kept it ringing with music when they were not running on inside the shed. They seemed to warble as easily as most birds breathe; in fact, song seemed a necessity to them. There was a high pole in front of the shed, and one day I found my ebullient little friend squatting on top to hold himself on while he sang out at the top of his lungs! Another time I came face to face with a pair when the songster was in the midst of his roundelay. He stopped short, bobbed nervously from side to side, and then, rising to his feet and putting his right foot forward with a pretty courageous gesture, took up his song again. When the pair were building in the crate, I stuck some white hen's feathers there, thinking they might like to use them. Mr. Troglodytes came first, and seeing them, instead of turning tail as I have known brave guardians of the nest to do, burst out singing, as if it were a huge joke. Then he hopped down on the rim of the box to scrutinize the plumes, after which he flew out. But he had to stop to sing atilt of an elder stem before he could go on to tell his spouse about them.

One day, when riding back to the ranch, I saw half a dozen turkey buzzards soaring over the meadow — perhaps there was a dead jack-rabbit in the field. It was astonishing to see how soon the birds would discover small carrion from their great height. The ranchman never thought of burying anything, they were such good scavengers. A few hours after an animal was thrown out in the field the vultures would find it. They would stand on the body and pull it to pieces in the most revolting way, The ranchman told me he had seen them circle over a pair of fighting snakes, waiting to devour the one that was injured. They were grotesque birds. I often saw them walk with their wings held out at their sides as if cooling themselves, and the unbird-like attitude together with the horrid appearance of their red skinny heads made them seem more like harpies than before.

They were most interesting at a distance. I once saw three of them standing like black images on a granite bowlder, on top of a hill overlooking the valley. After a moment they set out and went circling in the sky. Although they flew in a group, it seemed as if the individual birds respected one another's lines so as not to cover the same ground. Sometimes when soaring they seemed to rest on the air and let themselves be borne by the wind; for they wobbled from one side to the other like a cork on rough water.

One of the most interesting birds of the valley is the road-runner or chaparral cock, a grayish brown bird who stands almost as high as a crow and has a tail as long as a magpie's. He is noted for his swiftness of foot. Sometimes, when we were driving over the hills, a road-runner would start out of the brush on a lonely part of the road and for quite a distance keep ahead of the horses, although they trotted freely along. When tired of running he would dash off into the brush, where he stopped himself by suddenly throwing his long tail over his back. A Texan, in talking of the bird, said, "It takes a right peart cur to catch one," and added that when a road-runner is chased he will rise but once, for his main reliance

Valley Quail and Road-Runner.

is in his running, and he does not trust much to his short wings. The chaparral cocks nested in the cactus on our hills, and were said to live largely on lizards and horned toads.

It became evident that a pair of these singular birds had taken up quarters in the chaparral on the hillside back of the ranch-house, for one of them was often seen with the hens in the door-yard. One day I was talking to the ranchman when the road-runner appeared. He paid no attention to us, but went straight to the hen-house, apparently to get cocoons. Looking between the laths, I could see him at work. He flew up on the hen-roosts as if quite at home; he had been there before and knew the ways of the house. He even dashed into the peak of the roof and brought down the white cocoon balls dangling with cobweb. When he had finished his hunt he stood in the doorway, and a pair of blackbirds lit on the fence post over his head, looking down at him wonderingly. Was he a new kind of hen? He was almost as big as a bantam. They sat and looked at him, and he stood and stared at them till all three were satisfied, when the blackbirds flew off and the road-runner walked out by the kitchen to hunt among the buckets for food.

These curious birds seem to be of an inquiring turn of mind, and sometimes their investigations end sadly. The windmills, which are a new thing in this dry land, naturally stimulate their curiosity. A small boy from the neighboring town —Escondido—told me that he had known four road-runners to get drowned in one tank; though he corrected himself afterwards by saying, "We fished out *one* before he got drowned!"

Another lad told me he had seen road-runners

in the nesting season call for their mates on the hills. He had seen one stand on a bowlder fifteen feet high, and after strutting up and down the rock with his tail and wings hanging, stop to call, putting his bill down on the rock and going through contortions as if pumping out the sound. The lad thought his calls were answered from the brush below.

In April the ranchman reported that he had seen dusky poor-wills, relatives of our whip-poor-wills, out flycatching on the road beyond the ranch-house after dark. He had seen as many as eight or nine at once, and they had let him come within three feet of them. Accordingly, one night right after tea I started out to see them. The poor-wills choose the most beautiful part of the twenty-four hours for their activity. When I went out, the sky above the dark wall of the valley was a quiet greenish yellow, and the rosy light was fading in the north at the head of the canyon. White masses of fog pushed in from the ocean. Then the constellations dawned and brightened till the evening star shone out in her full radiant beauty. Locusts and crickets droned; bats zigzagged overhead; and suddenly from the dusty road some black objects started up, fluttered low over the barley, and dropped back on the road again. At the same time came the call of the poor-will, which, close at hand, is a soft burring *poor-will*, *poor-wil'-low*. Two or three hours later I went out again. The full moon had risen, and shone down, transforming the landscape. The road was a narrow line between silvered fields of headed grain, and the granite bowlders gleamed white on the hills inclosing the sleeping valley. For a few moments the shrill barking of coyote wolves disturbed the stillness; then again the night became silent; peace rested upon the valley, and from far up the canyon came the faint, sad cry, *poor-wil'-low*, *poor-wil'-low*.

VIII.

POCKET MAKERS.

THE bush-tits are cousins of the eastern chickadees, which is reason enough for liking them, although the California fruit growers have a more substantial reason in the way the birds eat the scale that injures the olive-trees. The bush-tits might be the little sisters of the chickadee family, they are so small. They look like gray balls with long tails attached, for they are plump fluffy tots, no bigger than your thumb, without their tails. One of them, when preoccupied, once came within three feet of where I stood. When he discovered me a comical look of surprise came into his yellow eyes and he went tilting off, for his long tail gave him a pitching flight as if he were about to go on his bill, a flight that reminds one of the tail that wagged the dog.

There were so many of the gray pocket nests in the oaks that it was hard to choose which to watch, but one of the most interesting hung from a branch of the big double oak of the gnatcatchers, above the ranch-house, where I could see it when sitting in the crotch of the tree. While watching it I looked beyond over the chap-

Nest of the Bush-tit.

arral wall away to a dark purple peak standing against a sky flecked with sun-whitened clouds. The nest was like an oriole's, but nearly twice as long, though the builders were less than half the size of the orioles. Instead of being open at the top, it was roofed over, and the only entrance was a small round hole, the girth of the bird, about two inches under the roof.

One might imagine that such big houses would be dark with only one small dormer window, and the valley children assured me that the birds hung living firefly lamps on their walls! I suggested that a Society for the Prevention of Cruelty to Fireflies would be needed if that were the case; but when it comes to that, what bird would choose to brood by gaslight?

When I first saw the bush-tit in its round doorway, it suggested Jack Horner's famous plum, comical little ball of feathers! When first watching the nest the small pair put me on their list of enemies, along with small boys, blue jays, and owls. To go down into the pocket under my stare seemed a terrible thing. When one of them came with a bit of moss for lining, it started for the front door, saw me, stopped, and turned to go to the back of the nest. Then it tried to get up courage to approach the house from the side, got in a panic and dashed against the wall as if expecting a door would open for it. When at last it did make bold to dart into the nest it was struck with terror, and, whisking around, jabbed the moss into the outside wall and fled!

Seeing that nothing awful happened, the birds finally took me off the black list and allowed me to oversee their work, as long as I gave no directions. Sometimes both little tots went down into the bag to work together; surely there was plenty of room for many such as they. But it is not always a matter of cubic inches, and one morning when the second bird was about to pop in, apparently it was advised to wait a minute. There was no ill feeling, though, for when the small builder came out it flew to the twig in front of the door, where its mate was waiting, and sat down beside it, a little Darby by his Joan.

They worked busily. Sometimes they popped in only to pop out again; at other times they stayed inside as long as if they had been human housekeepers, hanging pictures, straightening chairs, and setting their bric-a-brac in order for the fortieth time; each change requiring mature deliberation.

One morning — after the birds had been putting in lining long enough to have wadded half a dozen nests — if my judgment is of any value in such matters — I discovered that the roof was falling in; it was almost on top of the front door! The next day, to my dismay, the door had vanished. What was the trouble? Were the pretty pair young builders; was this their first nest, and had they paid more attention to decorating their house inside than to laying strong foundations; or had their pocket been too heavy for its frame?

However it came about, the wise birds concluded that they would not waste time crying over spilt milk. They calmly went to work to tear the first nest to pieces and build a second one out of it. One of them tweaked out its board with such a jerk it sent the pocket swinging like a pendulum. But the next time it wisely planted its claw firmly to steady itself, while it cautiously pulled the material out with its bill.

If the birds were inexperienced, they were bright enough to profit by experience. This time they hung their nest between the forks of a strong twig which had a cross twig to support the roof, so that the accident that had befallen them could not possibly occur again. They began work at the top, holding onto the twig with their claws and swinging themselves down inside to put in their material; and they moulded and shaped the pocket as they went along.

After watching the progress of the new nest, I went to see what had become of the old one. It was on the ground. On taking it home and pulling it to pieces, I found that the wall was from half an inch to an inch thick, made of fine gray moss and oak blossoms. There was a thick wadding of feathers inside. I counted *three hundred*, and there were a great many more! The amount of hard labor this stood for amazed me. No wonder the nest pulled down, with a whole featherbed inside! Why had they put it in? I asked some children, and one said, "To keep the eggs warm, I guess;" while the other suggested, "So the eggs would n't break." Most of the feathers were small, but there must have been several dozen chicken's feathers from two to three inches long. Among them was a plume of an owl.

Much to my surprise, in the bush-tit's nest there was a broken eggshell. Had the egg broken in falling, or had a snake been there? One of the boys of the valley told me about seeing a racer snake go into a bush-tit's pocket. The cries of the birds rallied several other pairs, and they all flew about in distress, though not one of them dared touch the dreadful tail that hung out of the nest hole. As the snake was about three feet long, the pocket bulged as it moved around inside. There were four nestlings about a quarter grown, and the relentless creature devoured them all. The boy waited below with a stick, and when it came out, killed it and shook it by the tail till the small birds popped out of its mouth. If my broken eggshell pointed to any such tragedy, it cleared the birds of the accusation of being poor builders.

The nest, which the first day was a filmy spot in the leaves, by the next day had become a gray pocket over eight inches long, although I could still see daylight through it. In working, the birds flew to the top of the open bag and hopped down inside. I could see the pocket shake and bulge as they worked within. When

POCKET NEST IN AN OAK

they flew away to any distance, on their return they almost always came with their little call of *schrit, schrit.*

This nest was so low that I used to throw myself on the sand beneath the tree to watch it, taking many a sunbath there, with hat drawn down till I could just see the nest in the pendent branches, and watch the changing mosaics made by the sky through the moving leaves. When resting on the sand the thought of rattlesnakes came to me, for the brush on either side was a shelter for them, and they might easily have crept up beside me without my hearing them.

The second bush-tit's nest was shorter than the first one. Perhaps the builders thought the length had something to do with the fall of the first; or perhaps they did n't feel like collecting three hundred more feathers, with oak blossoms and moss to match. They first put the frame of the front door below the supporting cross twig, and then, as if they thought it needed more support, changed it and put the door above the twig, so that the roof could not possibly close the hole, even if it did fall in. The doorway was also made much larger than that of the first nest.

After making away with the old nest, my conscience smote me. Perhaps the little pocket makers were not through with it, even if it was on the ground; so I brought a piece of it back and tied it with a grass stem to a twig below the nest they were at work on, to save them as much trouble as might be. When my bird came, her bright eyes were quick to espy the old nest. She looked around, bewildered, as if wondering whether she was really awake, and making sure that this strange looking affair were not her second nest, come to grief in her absence. Being reassured by her examination, she came back and hopped from twig to twig inspecting the old piece of nest. At last she caught sight of a feather. That, apparently, was just what she wanted. She quickly flew over, pulled out the white plume, and went straight to the new house with it!

I was not able to watch any of my bush-tits through the season, that year, but five years later, when again in southern California, to my delight I found the tits building in almost the same tree where they had been before.

One day an interesting brood was out in the brush, and I took notes on their proceedings: "A family of young were abroad this morning filling the leaves with their little moving forms, and the air with their fledgling cry of *schrit*. As nearly as I could judge, there were ten in the family — eight young tagging after two old birds. While I watched, a droll thing happened, proving that a family of eight may affect a parent's breakfast as well as his nerves. One of the family, which I took to be the father bird, had some goody in his bill, and one of the young, presumably, followed him for it, flying up on his twig. The old bird turned his back upon the little one and went on shaking the grub. Presently a second one flew down on the other side of him, — he was between two fires; they touched him on both sides. I watched with interest to see what he would do about it, and was much amused when he opened his wings and flew up over their heads out of reach! Would he come back to feed them after his food was properly prepared? No, — he sat up on the branch and ate the morsel himself! I was rather shocked by such a deliberate proceeding, but then it occurred to me that parent birds have to take a bite themselves once in a while; though of course their business is to feed the children!"

IX.

THE BIG SYCAMORE.

BEFORE going home from my morning sessions with the little lover and other feathered friends, I often took a gallop at the foot of the hills to visit a gigantic old tree, the king of the valley. One such ride is especially marked in my memory. It was on one of California's most perfect mornings. When the sun had risen over the valley, the fog dissolved before it, sinking away until only small white clouds were left in the tender blue of the notches between the red hills; while the bared vault overhead had that pure, deep, satisfying color peculiar to fog-cleared skies; and the cool fresh air was full of exhilaration. It put Mountain Billy so in tune with the morning that, when I chirrupped to him, shaking the reins on his neck, he quickly broke into a lope and his ringing hoofs beat time to my song as we sped down the valley, past vineyards and orchards and yellow fields of ripening grain. The free swift motion was a delight in itself, and after days and weeks given to the details of nest-making, shut away from the world in our little remote valley at the foot of the mountains, now, when we came to a break in the hills and our nostrils were greeted by the cool salt breeze coming from the Pacific, suddenly the whole horizon broadened; the inclosing valley walls were overlooked; we were galloping under the high arching heavens in a wind blowing from far over the wide ocean.

Here stood the great sycamore, with branches swaying; for the tree faced this break in the hills. It seemed as if the old monarch, with roots firmly planted, had battled for its ground; and now, as a conqueror, stood with arms uplifted to meet the ocean gales. I had never before appreciated the dignity of those straight upreared shafts, the vital strength of those deep grappling roots, the mighty grandeur of this old battle king.

When one of the trunks fell, I had to hunt the sycamore over to find where it came from, not missing it in the massive framework that was left. The giant measured twenty-three feet and a half in circumference, three feet from the ground. Its enormous branches stretched out horizontally so far that, between the body of the tree and the tips that hung to the earth, there was a wide corridor where one could promenade on horseback. In fact, the tree spanned, from the tip of one branch to the tip of the other, one hundred and fifty-eight feet. In the photograph, the figure of a person is almost lost in the complicated network of the frame of the tree. The treetop was a grove in itself. A flock of blackbirds flying up into it was lost among the branches.

The ranchman knew the sycamore as the 'swallow tree,' because in former years, before the valley was settled, swallows that have since taken to barns built there. Between three and four hundred of them plastered their nests on the underside of the big limbs, about half way up the tree, where the bark was rough. They built so close together that the nests made a solid mass of mud. For several seasons, it was said, "they had bad luck." They began building before the rainy season was over, and all but a few dozen nests which were in especially protected places were swept away. The number of nests was so enormous that the ground was covered several inches deep with mud.

Billy used to improve his time by nibbling barley while I watched birds in the sycamore corridor. We had not been there long before I discovered a bee's nest in the hollow of one of the trunks. The owners were busily flying in and out, and a pair of big bee-birds flew down from their nest in the treetop and saved themselves trouble by lunching at this convenient ground floor restaurant. As I sat on Billy, facing the nest, one of the pair swept down over the mouth of the hole, caught a bee and settled back on the branch to swallow it. This seemed to be the regular performance, and was kept up so

THE BIG SYCAMORE

continuously, even when we were standing close by, that if, as is supposed, the birds eat only drones, few but workers would be left in that hive.

The flycatchers seemed well suited to the sycamore; they were birds of large ideas and sweeping flights. Their nest was at the top of the tree, probably eighty feet from the ground, but when one of them flew down, instead of coming a branch at a time, he would set his wings and, giving a loud cry, — as a child shouts when pushing off his sled at the top of a steep hill, — he would sail obliquely down from the treetop to the foot of the hillside beyond. When looking for his material he would hover over the field like a phœbe. Then, on returning, unlike the other birds who lived in the tree and used the branches as ladders, he would start from the ground and with labored flights climb obliquely up the air to the treetop. Once his material dangled a foot behind him. The birds seemed to enjoy these great flights.

Their nest was not finished, and while one went for material, the other — presumably the male — guarded the nest. As there was nothing to guard as yet, it often seemed a matter of venting his own spleen! When not occupied in arranging his plumes, he would shoot down at every small bird that came upstairs; a cowardly proceeding, but perhaps he thought it necessary to keep his hand in against meeting bigger boys than he! When coming with material, one of the bee-birds got caught in a heavy rope of cobweb that dangled from the nest, and had to flutter hard to extricate itself. About their nests these birds seemed as home-loving as any others. Their domesticity quite surprised me; they had always seemed such harsh, scolding, aggressive birds! When one of them sat among the green leaves, pluming the soft sulphur yellow feathers of its breast, it looked so gentle and attractive that it was a shock when the familiar petulant screams again jarred the air. The birds often hunted from the fence beyond the sycamore, and flew from post to post with legs dangling, shaking their wings as they lit, with a shrill *kit' r' r' r' r'.*

The sycamore was a regular apartment house; so many birds were moving among the boughs it was impossible to tell where they all lived. One day I found a pair of doves sitting on a sunny branch above me. The one I took to be the male sat perched crosswise, while his mate sat facing him, lengthwise of the limb. He calmly fluffed out his feathers and preened himself, while his meek spouse watched him. She fluttered her wings, teasing him to feed her, but he kept on dressing out his plumes. Then she edged a little closer, and almost essayed to touch his majesty with her pretty blue bill, but he sat with lordly composure quite ignoring her existence till a blackbird bustled up, when they both started nervously, and turning, sat demurely side by side on the limb, the wind tilting their long tails.

A pair of bright orange orioles had a nest in the sycamore, though I never should have known it had I not seen them go to it to feed their young. It was a well shaded cradle surely, with its canopy of big green leaves.

There were a good many hints to be had, first and last. A song sparrow appeared and stood on a branch with its tail perked up in a business-like way as if it had been feeding a brood. A wren came to the tree, — a mere pinch of feathers in the giant sycamore, — and though I lost sight of it, many a hollow up in the fourteenth story might have afforded a home for the pretty dear without any one's being the wiser, unless it were the bee-bird in the attic. A family of bush-tits flew about in the sycamore top, looking like pin-heads in a grove of trees. A black phœbe sometimes lit on the fence posts under the branches — it wanted to find a nesting place about the windmill in the opposite field, I felt sure, though a boy had told me that the bird sometimes plastered its nest onto the branches of the big tree itself. Besides all the rest, rosy linnets and blue lazuli buntings made the old tree ring with their musical roundelays.

One day when I rode down to the sycamore,

the meadow bordering it was full of haycocks, and a rabbit ran out from under one of them, frightened by the clatter of Billy's hoofs. That morning the tree was fairly alive with blackbirds and doves — what a deafening medley the blackbirds made! In the fields near the sycamore flocks of redwings went swinging over the tall gleaming mustard. This was a great place for blackbirds, for the big tree was on the edge of the one piece of marsh land in the valley, and they were quick to take advantage of its reeds for nesting places.

The cienaga — as they called the swamp — was used as a pasture. It was pleasant to look out upon, from under the branches of the great tree. A group of horses stood in the shade of a cluster of oaks on the farther side of it, while the cows, a beautiful herd of buff and white Guernseys, waded through the swamp grass to drink near the sycamore, and the blackbirds wound in and out among them. I had been in a dry land so long it was hard to believe there was actual water in the marsh till I saw it drip from their chins and heard the sucking sound as they laboriously dragged their feet out of the mud — a noise that took me back to eastern pastures, but sounded strangely unfamiliar here in this rainless land. One of the pretty Guernseys with a white star in her forehead strayed up under the tree, and the shadows of the leaves moved over her as she raised her sensitive face to see who was there.

The son of the ranchman who owned the dairy — the one who invited me down to see the play between his dog Romulus and the burrowing owl — said that when herding cows by the sycamore he once caught sight of a coyote wolf. He clapped his hands to send his dog, Romulus, after the wolf; and the noise frightened the wild creature so that he started to run up the hill across the road from the sycamore. Romulus followed hard at his heels till they got well up the hillside, when the coyote felt that he was on his own ground and turned on the dog, who fled back to his master with his tail between his legs. The lad, clapping his hands, set the dog on the coyote again, and this animated but bloodless performance was repeated and kept up till both were tired out, the animals chasing each other back and forth from the sycamore to the hillside with as much energy and perhaps as much courage as was displayed by that historic king of France who had five thousand men and —

> "... marched them up a hill and then
> He marched them down again."

On one side of the sycamore was a great wall of weeds higher than my head when on horseback; a dense mass of yellow mustard, and fragrant wild celery which was covered with delicate white bloom. I saw blackbirds carrying material into this thicket, but as I had known of neighbors' horses getting bitten by rattlesnakes among the high weeds, did not think it worth while to wade around in it much for such common birds as they. But one day, seeing a pair of rare blue grosbeaks fly down into the tangle, I turned Billy right in after them, though holding his head well up in consideration of the snakes. The birds vanished, so we stood still to wait. Suddenly I heard a slight sound as of something slipping through the weeds at Billy's feet, and looking down saw a snake marked like a rattler; and as it slid by Billy's hoof I noticed with horror that the end of its tail was blunt — the harmless gopher snake that resembles the rattler has a tapering tail! I gazed at it spellbound, but in the dim light could not make out whether it had rattles or not. I had seen enough, however, and whipping up Billy was out of those weeds in a hurry. Safely outside, I looked at my little horse remorsefully — what if my desire to see a new nest had been the cause of his getting a rattlesnake bite!

The next day when I went down to the sycamore a German was mowing there with a pair of mules. He was a typical Rhinelander, with blue eyes and long curling hair and beard, and as he drove he sang in a deep rich voice one of the beautiful melodies of his fatherland.

Screened by the branches, I listened quite unmindful of my work till my reverie was interrupted by the man's giving a harsh cry to his mules. It was only an aside, however, for he dropped back into his song in the same rich sympathetic voice.

In riding out from the tree on my way home, I saw that he was mowing just where the snake had been, and warned him to be careful lest the horses get bitten. At the word rattlesnake his blue eyes dilated, and he assured me that he would be on his guard. Seeing my glasses and note-book, he asked if I were studying birds. When told that I was, from his seat on the mowing-machine he took off his hat and bowed with the air of a lord, saying in broken English, "I am pleased to meet you!" — a pleasant tribute to the profession. A few days later, on meeting him, he asked if I had found the rattlesnake — he had killed it under the sycamore and hung it on a branch for me to see.

As the memory of my morning rides down to the sycamore brings to mind the wonderful freshness of California's fog-cleared skies, so my sunset rides home from the great tree recall the peacefulness of the quiet valley at twilight. One sunset stands out with peculiar distinctness. As Mountain Billy turned from the sycamore marsh its leaning blades gleamed in the evening light, and the sun warmed the sides of the line of buff

Guernseys wading in procession through the high swamp grass to their out-door milking stand. Beyond, a load of hay was crossing the meadows with sun on the reins and the pitchforks the men carried over their shoulders; and beyond, at the head of the valley, the western canyons were filled with golden haze, while the last shafts of yellow light loitered over the apricot orchards below, where the tranquil birds were singing their evening songs. Slowly the long shadows of the mountain crept over orchard and vineyard until, finally, the sun rounded the last peak and left our little valley in darkness.

X.

AMONG MY TENANTS.

THE first year I was in California the thought of the orchards that were to be set out on my ranch appealed to me much less than what the place already possessed. As an inheritance from the stream that came down in spring through the Ughland canyon — past the homes of the little lover, the gnatcatchers, the little prisoners, and the lazulis and blue jays — there was a straggling line of old sycamores, full of birds' nests ; and a patch of weeds, wild mustard, and willows, which was a capital shelter for wandering warblers; and a bright sunny spot always ringing with songs.

So many houses were being put up without so much as a by-your-leave that it was high time for an ornithological landlady to bestir herself and look to her ornithological squatters ; so, day after day I turned my horse toward the ranch and spent the morning getting acquainted with my tenants, riding along the shady line and making friendly calls at each tree.

Half of the blackbirds who worked in the vineyard must have been beholden to me for rent, I should judge by the jolly choruses of the sable hordes moving about my treetops. There was a bee's nest in one of the sycamores, and one day the buzzing mob 'took after me' so madly that I had to whip up Canello and beat about with my hat to get clear of them.

Another day, when we stopped under a sycamore, such a loud shrill whistle sounded suddenly overhead that the horse started. A big bird in black sat with feathers bristled up about him like a threatening raven, croaking away sepulchrally directly overhead, bending down gazing at us out of his yellow eyes as if to see how we took it. It was a laughable sight. Blackbirds seem such human, humorous birds one can almost fancy them playing such pranks just for the fun of it.

The blackbird colony was a busy one nesting-time. The builders would fly down to the road to get material, stepping along quickly, looking from side to side with an alert, business-like air, as if they knew just what they wanted. Some of them used the button-balls to line their nests.

A pair had built in one of the round mats of mistletoe at the end of a branch, and while looking at the nest one day I was amazed to see a butcherbird come flying in a straight line toward it. He did not reach his destination, for while still in air both blackbirds darted down at him and drove him back faster than he had come. The guardian of the nest escorted him almost home, and when the victorious pair were returning

ALONG THE LINE OF SYCAMORES

they were joined by a noisy band of indignant members of the blackbird clan.

I watched this attack with great interest, not knowing that shrikes were concerned in blackbird matters, and also because it was welcome news that one of these strange characters had rented a lot of me. I made a note of the direction my outlaw tenant took when driven ignominiously home, and at my earliest convenience called. Such cruel tales are told of his cold-blooded way of impaling birds and beasts upon thorns and barbed wires that one naturally looks upon him as a monster; but I found that he, like many another villain, turns a gentle face to his nest.

He had pitched his tent on the farthest outpost of my ranch in a little bunch of willows, weeds, and mustard — long since converted into a well-kept prune orchard. The nest, which was a big round mass of sticks, was inside the willows in a clump of dry stalks about six feet from the ground. I had hardly found it before one of the builders swooped down to it right before my eyes, with the hardihood of one who fears no man; though it must be acknowledged that the shrikes, like other birds on the ranch, were so used to grazing horses they quite naturally took me for a cattle herder.

In this case Canello did not act as my ally. He had been quiet and docile most of the morning, but now was hungry and saw some grass he was bent on having, so took the bit in his teeth and made such an obstinate fight that, before I had conquered him, the shrikes had left the premises and my call was finished without my hosts.

On my next visit Canello behaved in more seemly manner, and permitted me to see something of the ways of the maligned birds. You would not have known them from any one else except for the remarkable stillness of their neighborhood. Some finches flew overhead as if meaning to stop, but saw the shrike and went on. I could hear the merry songs of the assembly down in the sycamores, but not a bird lit while we were there — the shrikes certainly have a bad name among their neighbors. They had a proud bearing and an imperative manner, but seemed so gentle and human in their domestic life that my prejudices were softened, as one's generally are by near acquaintance, and I became really very fond of my handsome tenants.

It looked as if the shrike fed his mate. At any rate, they worked together and rested together, perching in lordly fashion high on the willows overlooking their home. They did not object to observers when at work. One day, when Canello's nose appeared by the nest, the builder looked at him over her shoulder and then quietly slid off the nest, flying up on her perch to wait till he should leave. It was a temptation to keep her waiting some time, for the shrike's corner was a pleasant place to linger in. The sea-breeze was so strong it turned the willow leaves white side out, and the beautiful glistening mustard grew so high there that when Canello walked into it, the golden blossoms waved over our heads. We haunted the premises till the birds had finished their framework, put in a lining of snow-white plant cotton, and had laid four eggs.

But when getting to feel like an old friend of the family, on riding down one day I found the nest lying in the dust of the road broken and despoiled. It made me as unhappy as if the outlaws had been unimpeachable bird citizens — which comes of knowing both sides of a person's character! Do birds hand down traditions of ill luck? However it may be, five years later I found the nest of a pair in a dark mat of mistletoe at the end of a high oak branch, which was a much safer place than the low willow.

While I was watching the first shrike family, Canello had two scares. Once when we were standing still by the willow we heard what sounded like a rattlesnake springing its rattle. The nervous horse pricked up his ears, raised his head, and looked in the grass as if he saw snakes, and though I succeeded in quieting him, when we went home he started at every stick and was ready to shy at every shadow. Another morning he saw a Mexican riding along by the vineyard, a man with a very dark face and a red shirt.

Canello acted much as he had when hearing the rattlesnake, and did not quiet down till horse and rider were out of sight. The ranch-man told me he had been cruelly treated by the Mexican who broke him, so perhaps it was another case of association of ideas.

East of the willows, and separated from them by the dark green mallows and bright yellow California forget-me-nots, was the sycamore where the shrike was driven off by the blackbirds. Here a little brown wren had taken up her abode. The nest was in a dead limb with a lengthwise slit, and a scoop at the end like an apple-corer, so when one of the wrens flew down its hole with a stick, the twig stuck out of the crack as she ran along with it. She quite won my heart by her frank way of meeting her landlady. Instead of flying off, she looked me over and then quietly sat down in her doorway to wait for her mate.

On the road to my sycamores was a deserted whitewashed adobe. The place had become overgrown with weeds, vines, and bushes, and was taken possession of by squirrels and birds. Nature had reclaimed it, covering its ugly scars with garlands, and making it bloom under her tender touch. One morning, as I rode by, a black phœbe was perched on the old adobe chimney of the little house, while his mate sat on the board that covered the well, in a way that made it easy to jump to a conclusion. When she flew up to the acacia beside the well and looked down anxiously, I put the pair on my calling list. It did not take many visits to prove my conclusion — there was a nest down in the well with white eggs in it. The phœbes were most trustful birds, and not only let Canello tramp around their yard, but when a pump was put down the well, and water pumped up day by day, the brave parents, instead of deserting their eggs, went on brooding as if nothing had happened.

Black Phœbe.
(One half natural size.)

Five years later, on going back to the ranch, I found the phœbes around the old place, but hunted in vain for the nest. A schoolhouse had been built in the interval, near the old adobe, and the birds perched on its gables, on the hitching posts in front of it, and on my prune-trees, that had taken the place of the willows, across the road. They even came up to my small ranch-house and filled me with delightful anticipations by inspecting the beams of the piazza; but they could not find what they wanted and flew off to build elsewhere.

Eastern Phœbe.
(One half natural size.)

Later in the season, a neighbor whose ranch was opposite mine showed me a phœbe's nest inside his whitewashed chicken house. It was a mud pocket like a swallow's, made of large pellets of mud plastered against a board in the peak of the house. Of course I could never prove that these birds were my old friends, but it seemed very probable.

The smallest of my tenants was a humming-bird. I saw it fly into a low spray, and it stayed there so long that when it left I rode up to look, and found that it was building on the tip of a twig under a sycamore leaf umbrella, one whose veining showed against the light. By rising in the saddle I could just reach the twig and pull it down to look inside the nest; but afterwards I found so many other hummers who could be watched with fewer gymnastics, I rested content with knowing that this little friend was there.

One morning, when on the way to the sycamores, I found an oriole's nest high in a tree. Canello was hungry, but when permitted to eat barley under the branches kept reasonably quiet. There were two species of orioles in the valley; and not knowing to which the nest belonged, I prepared to wait for the return of the owner. The heat was so oppressive that I took off my hat, and a bird flew into the tree with bill open, gasping. After my hot ride down the valley the shade of the big tree was very grateful; and the cool trade wind coming through a gap in the hills most refreshing.

Suddenly there was a flash — we all waked up — was that the house owner? What a remarkable bird! and what a display of color! — it had a red head, fiery in the sun; a black back, and a vivid yellow breast. On looking it up in Ridgway the stranger proved to be the Louisiana tanager, a high mountain bird. That was a red letter day for me. No one can know, without experiencing it, the delight of such discoveries. The pleasure is as genuine as if the world were made anew for you. In the excitement the oriole's nest was neglected; but ordinarily the rare unknown birds did not detract from the enjoyment of the old, more familiar ones.

So when the brilliant stranger flew away and was seen no more I turned with pleasure to the pair of sparrow hawks who had come to live on the ranch. A branch had fallen from one of the trees, and the hawks found its hollow just suited to their needs. It was a good, spacious house, but a pair of their cousins who had built in a tree over the whitewashed hovel had made a sad mistake in choosing their dwelling — for the front door was so small they could hardly enter! I used to stop to watch them, and was very much amused at their efforts to make the best of it.

Canello could stand up to his knees in alfilaree

clover under their tree, so he allowed me to watch the birds in peace. The first day the male sparrow hawk flew to the tree with what looked like a snake dangling from his bill, and as he alighted screamed *kit-kit'ar' r' r' r'*, spreading his wings and shaking them with emphasis. When this brought no response, he flew from branch to branch, crying out lustily. He revolved around the end of a broken limb in whose small hollow was framed the head of Madame Falco. From her height she looked like a rag doll at her window. Her funny round face, which filled the doorway, had black spots for bill and eyes, and dark lines down the cheeks that might have simulated rag doll tattooing.

Evidently there was some reason why she did not want to come to breakfast. Once she started to turn back into the nest, but at last laboriously wedged her way out of the hole and flew to a branch. Her mate was at her side in an instant, and handed her the snake. She took it greedily and flew off with it, let us hope because she was afraid of me, not because she did not want to divide with him, or thought he would ask her to, after all his devotion and patience!

When the bird went back to her nest, her hesitation about leaving it was explained. For a long time she sat on a limb near by with tail bobbing, apparently trying to make up her mind to go in. When she did fly up at the hole she could not get in, and half fell down. After this failure she sat down on a branch, her tail tilting as violently as a pipit's, and when Canello moved around too much, took the excuse and flew off. Her mate came back with her, but when he saw us, he screamed and flew away, leaving her to her fate.

She sat looking at her hole a long time before she tried it again, and when she did try, failed. It was not till her fourth attempt that she succeeded. The hole was very much too small for her, and the surface of the branch below it was so smooth and slippery that it gave her nothing to hold to in trying to wedge herself in. She would fly against the hole and attempt to hook her bill over the edge, and so draw herself up, but her shoulders were too big for the space. She tried to make them smaller by drawing down her wings lengthwise. Once, in her efforts, she spread her tail like a fan. After her third struggle, she sat for a long time smoothing her ruffled feathers, shaking herself, scratching her face with her foot and trying to get her plumes in order.

While making her toilet she apparently thought of a new plan. She went back to the hole and, raising her claw, fastened it inside the hole and with a spasmodic effort wedged in her body and disappeared down the black hollow.

Her mate came a moment after, but she did not even appear in the doorway when he called. Again he came, crying *keek' keek' kick'-er' r' r'*, in tender falsetto; but it was no use. Madame Falco had had altogether too hard a time getting in, to go out again in a hurry. He held a worm in his bill till he was tired, changed it to his claw, letting it dangle from that for a while; and then, as she would make no sign, finally flew off.

The next day we had another session with the sparrow hawk. She had evidently profited by experience. She did not fly at the hole in the violent way she had done the day before, but ambled along a limb to get as close to it as possible, and then quietly flew up. She made two or three unsuccessful attempts to enter, but kept at the branch, — falling back but once. She got half way in once or twice, but could not force her wings through. She acted as if determined not to give up, and at last, when she found herself falling backwards, with a desperate effort drew herself in.

There was another sparrow hawk family across the road from my ranch. In riding by one day, I saw a youngster looking out from the nest hole with big frightened eyes. Was it the only child, or was it monopolizing the fresh air while its brothers were smothering below? Another day there were two heads in the window; one was the round domed top of a fluffy nestling whose eyes expressed only vague fear; but the other was the strongly marked head of an old sparrow hawk, who eyed us with keen intelligence. As I stared up, the young one drew back into the hole behind its parent, probably in obedience to her command; and the old bird bent such an anxious inquiring gaze upon me that I took the hint and rode away to save the poor mother worry.

These were not the only hawks of the valley. Once, seeing one of the large Buteos winging its way with nesting sticks hanging from its claws, I turned Canello into the field after it, following till it lit in the top of a high sycamore. The pair were both gathering material. Sometimes they flew with the twigs in their claws; sometimes in their bills; now they would fly directly to the nest, again circle around the tree before alighting. When one was at work, the other sometimes flew up and soared so high in the sky he looked no larger than a sparrow hawk. In swooping to the ground suddenly, the hawks would hollow in their backs, stick up their tails, drop their legs for ballast, and so let themselves come to earth. While one of the birds was peacefully gathering sticks, two blackbirds attacked it, apparently on general grounds, because it belonged to a family that had been traduced since history began. To tell the honest truth, I trembled a little myself at thought of what might happen to some of my small tenants, though I reassured myself by remembering that the facts prove the maligned hawks much more likely to eat gophers than birds.

In the back of the stub occupied by one of the sparrow hawks it was a pleasure to find a flicker excavating its nest. Planting its claws firmly in the hole with tail braced against the bark, the bird leaned forward, thrusting its head in, over and again, as if feeding young. It used its feet as a pivot, and swung itself in, farther and farther, as it worked. Such gymnastics took strong feet, for the bird raised itself by them each time. It worked like an automatic toy wound up for the performance. When tired, the flicker hopped up on a branch and vented its feelings by shouting *if-if-if-if-if-if-if*, after which it quietly returned to work. The wood was so soft that the excavating made almost no noise, but it was easy to see what was going on, for the carpenter simply drew back its head and tossed out the glistening chips for all the world to see. At the end of a week the flicker was working so far down in its excavation that only the tip of its tail stuck out of the door.

The nest of another Colaptes, I found by accident — a fresh chip dropped from mid-air upon my riding skirt. Just then Canello gave a stentorian sneeze and the bird came to her window to look down. She did not object to us, and was loath to turn back inside the dark hole—such a close stuffy place—when outside there were the rich green leaves of the tree, the sweet breath of the hayfield and the gentle breeze just springing up; all the warmth and sunshine and fragrance of the fields. How could she ever leave to go below? Perhaps she bethought her that soon the dark hole would be a home ringing with the voices of her little ones; at all events, she quickly turned and disappeared in her nest.

At the foot of the ranch I discovered a comical, sleepy little brown owl, dozing in a sycamore window. When we waked it up, it went backing down the hole. I wondered if it kept awake all day without food, for surely owl children do not get many meals by daylight. I spoke to the ranchman's son about it, and he said he thought the old birds fed the young too much, that he had found about a dozen small kangaroo rats and mice in their holes! He told me that he had known old owls to change places in the daytime, and both birds to stay in the hole during the day. Down the valley, where an old well was only partly covered over, at different times he had found a number of drowned owls. They seemed to fly into any dark hole that offered. Three barn owls had been taken from a windmill tank in the neighborhood in about a month. In a mine at Escondido the man had found a number of owls sitting in a crevice where the earth had caved; and he had seen about a dozen of them fifty to a hundred feet underground, at the bottom of the mine shaft.

I did not wonder the birds wanted to keep out of sight in the daytime, knowing what happened to those that stayed out. A pair nested in the top of a high sycamore on my neighbors' premises, and when one stirred away from home, it did so to its sorrow. One morning there was such a commotion I rode down to see what was the matter. A big dark brown form flew down the avenue of sycamores ahead of us, followed by a mob of all the feathered house owners in the neighborhood. They escorted it home to the top of its own tree, where it seated itself on a limb, its big yellow eyes staring and its long ears dropped down, as if home were not home with a rout of angry bee-birds and blackbirds screeching and diving at you over your own doorsill. Two orioles started to fly over from the next tree, but went back, perhaps thinking it wiser not to make open war upon such near neighbors; while a sparrow hawk who came to help in the attack was judged too dangerous an ally and escorted home by a squad of blackbirds dispatched for the purpose. The poor persecuted owl screwed its head around to its back as if hoping to see pleasanter sights on that side; but the uncanny performance did not seem to please its enemies, and a blackbird flew rudely past, close under its bill, as if to warn it of what might happen.

The queerest of all my tenants was an old mother barn owl who lived in the black charred chimney of one of the sycamores. I found a white feather on the black wood one day in riding by, and pulling Canello up by the tree, broke off a twig and rapped on the door. She came blundering out and flew to a limb over our heads — such a queer old crone, with her hooked nose and her weazened face surrounded by a circlet of dark feathers. The light blinded her, and with her big round eyes wide open she leaned down staring to make out who we were. Then shaking her head reproachfully, she swayed solemnly from side to side. As the wind blew against her ragged feathers she drew her wings over her breast like a cloak, making herself look like a poverty-stricken wiseacre. Finding that we did not offer to go, the poor old crone took to her wings; but as she passed down the line of sycamores she roused the blackbird clan, and a pair of angry orioles flew out and attacked her. My conscience smote me for driving her out among her enemies, but on our return to the sycamores all was quiet again, and a lizard was sunning himself on the edge of the old owl's chimney.

XI.

AN UNNAMED BIRD.

Six years ago, on my first visit to California, I found a dainty cup of a nest out in the oaks, but the name of its owner was a puzzle. On returning East I consulted those who are wisest in matters of such fine china, but they were unable to clear up the matter. For five years that mystery haunted me. At the end of that time, when back in California, up in those same oaks, I found another cup of the same pattern; but the cup got broken and that was the end of it.

The fact of the matter is, you can identify perhaps ninety per cent. of the birds you see, with an opera-glass and — patience; but when it comes to the other ten per cent., including small vireos and flycatchers, and some others that might be mentioned, you are involved in perplexities that torment your mind and make you meditate murder; for it is impossible to

Name *all* the birds without a gun.

On bringing my riddle to the wise men, they shook their heads and asked why I did not shoot my bird and find out who he was. On saying the word his skin would be sent to me; but after knowing the little family in their home it would have been like raising my hand against familiar friends. Could I take their lives to gratify my curiosity about a name? I pondered long and weighed the matter well, trying to harden my heart; but the image of the winning trustful birds always rose before me and made it impossible. I will put the case before you, and you can judge if you would not have withheld your hand.

One day, hearing the sound of battle up in the treetops, I hurried over to the scene of action, when out dashed a pair of courageous little dull-colored birds in hot pursuit of a blue jay, whom they dove at till they drove him from the field. My sympathies were enlisted at once. Fearless little tots to brave a bird four times as big as themselves in defense of their home! How hard to have to build and rear a brood in the face of such a powerful foe! I wanted to take up the cudgels for them and stand guard to see that no harm came.

Planting my camp-stool under their oak, I watched eagerly to have my new friends show me their home. As I waited, a pair of turtle doves walked about on the sand under the farther branches of the tree; a pair of woodpeckers sat on a dead limb lying in wait for their prey; and a couple of titmice came hunting through the oak

— all the world seemed full of happy home-makers.

But soon I saw a sight that made me forget everything else. There were my brave little birds up in the oak working upon a beautiful moss cup that hung from a forked twig. They were building together, flying rapidly back and forth bringing bits of moss from the brush to put in their nest.

They worked independently, each hunting moss and placing it to its own satisfaction. What one did the other would be well pleased with, I felt sure. But while each worked according to its own ideas, they always appeared to be working together; they could not bear to be out of sight of each other long at a time. When the small father bird found himself at the nest alone, after placing his material he would stand and call to let his pretty mate know that he was waiting for her; or else sit down by the nest and warble over such a contented, happy little lay it warmed my heart just to listen to him.

When his mate appeared the merry birds would chase off for a race through the treetops. Song and play were mingled with their work, but, for all that, the happy builders' house grew under their hands, and they kept faithfully at their task of preparing the home for their little brood. Once the small, dainty mother bird, — surely it must have been she, — after putting in her bit of moss, settled down in the nest and sat there the picture of quiet happiness.

This was all I saw of the nest builders that year. A great storm swept through the valley, and it must have washed away the frail mossy cup, for it was gone and the tree was deserted. Nevertheless, the birds had been so attractive, and their nest so interesting, that through the five years that passed before my return to California I kept their memory green, and could never think of them without tenderness — though I could call them by no name. If they had only worn red feathers in their caps, it would have been some clue to their coats-of-arms; but, out of hand, there seemed to be nothing to mark the plain, little, greenish gray birds from half a dozen of their cousins.

When I finally returned to the California ranch, one of my first thoughts was for the moss nest makers up in the oaks. Now I had a chance to solve the mystery without harming one of their pretty feathers, for by long and patient watching I might get near enough to puzzle out the 'spurious primary' and the subtle distinctions of tint that make such a difference in calling birds by their right names.

For six weeks I watched and listened in vain, but one day when riding up the canyon rejoicing at the new life that filled the trees, I stopped under an oak only a few rods from the one where the nest had been five years before, and looking up saw a small dull-colored bird with a bit of moss in its bill walking down into a mossy cup right before my eyes! For a few moments I was the happiest observer in the land. I had found my little friend again, after all these years! It looked over the edge of the twig at me several times, but went on gathering material as unconcernedly as if it, too, remembered me. The mossy cup seemed prettier than any rare bit of Sèvres china, for I looked upon it with eyes that had been waiting for the sight for five years.

As the bird worked, a cottontail rabbit rustled the leaves, and Billy started forward, frightening the timid animal so that it scampered off over the ground, showing the white underside of its tail. But though Billy and the rabbit were both terrified, the brave worker only flew down to a twig to look at them, and turned back calmly to its task.

The nest was so protectively colored that I could not see it readily, and sometimes started to find that I had been looking right at it without knowing it. The prospect of identifying my birds was not encouraging. You might as well expect to see from the first floor what was going on up in a cupola as to expect to see from the ground what birds are doing up in the thick oak tops. You have reason to be thankful for even a glimpse of a bird in the heavy foliage, and as for 'spurious primaries,' — "Woe worth the chase!"

Now and then I got a hint of family matters. My two little friends were working together, and occasionally I saw a bit of moss put in; but it was evident that the main part of the work was over. One day I waited half an hour, and when the bird came it acted as if it had really done all that was necessary, and only returned for the sake of being about its pretty home.

The birds said a good deal up in the oak, sometimes in sweet lisping tones, as though talking to themselves about the nest. They often flew away from it not far over my head. The call note was a loud whistle — *whee-it'* — and the bird gave it so rapidly that I once took out my watch to time him, after which he called seventy times in sixty seconds. Often after whistling loudly he would give a soft low call. His clear ringing voice was one of the most cheering in the valley.

When the building seemed done and I was looking forward to the brooding, as the birds would then, perforce, be more about the nest, one sad morning I rode up through the oaks and found the beautiful moss cup torn and dangling from its branch. It was the keenest disappointment of the nesting season, and there had been many. The pretty acquaintance to whose renewal I had looked forward so many years was now ended.

Again I had to leave California without being able to name my winning little friends. If I had been too much interested in them before to set a price on their heads; now, rather than raise my voice against them, they should remain forever unnamed.[1]

[1] Since this paper was written, I have consulted an authority on nests, who thinks that this nameless bird was probably Hutton's vireo.

XII.

HUMMERS.

CALIFORNIA is the land of flowers and hummingbirds. Hummingbirds are there the winged companions of the flowers. In the valleys the airy birds hover about the filmy golden mustard and the sweet-scented primroses; on the blooming hillsides in spring the air is filled with whirring wings and piping voices, as the fairy troops pass and repass at their mad gambols. At one moment the birds are circling methodically around the whorls of the blue sage; at the next, hurtling through the air after a distant companion. The great wild gooseberry bushes with red fuchsia-like flowers are like bee-hives, swarming with noisy hummers. The whizzing and whirring lead one to the bushes from a distance, and on approaching one is met by the brown spindle-like birds, darting out from the blooming shrubs, gleams of green, gold, and scarlet glancing from their gorgets.

The large brown hummers probably stop in the valley only on their way north, but the little black-chinned ones make their home there, and the big spreading sycamores and the great live-oaks are their nesting grounds. In the big oak beside the ranch-house I have seen two or three nests at once; and a ring of live-oaks in front

The Little Hummer on her Bow-Knot Nest.
(From a photograph.)

of the house held a complement of nests. From the hammock under the oak beside the house one could watch the birds at their work. If the front door was left open, the hummers would sometimes fly inside; and as we stepped out they often darted away from the flowers growing under the windows.

California is the place of all places to study hummingbirds. The only drawback is that there are always too many other birds to watch at the same time; but one sees enough to want to see more. I never saw a hummingbird courtship unless — perhaps one performance I saw was part of the wooing. I was sitting on Mountain Billy under the little lover's sycamore when a buzzing and a whirring sounded overhead. On a twig sat a wee green lady and before her was her lover (?), who, with the sound and regularity of a spindle in a machine, swung shuttling from side to side in an arc less than a yard long. He never turned around, or took his eyes off his lady's, but threw himself back at the end of his line by a quick spread of his tail. She sat with her eyes fixed upon him, and as he moved from side to side her long bill followed him in a very droll way. When through with his dance he looked at her intently, as if to see what effect his performance had had upon her. She made some remark, apparently not to his liking, for when he had answered he flew away. She called after him, but as he did not return she stretched herself and flew up on a twig above with an amusing air of relief.

This is all I have ever seen of the courtship; but when it comes to nest-building, I have often been an eye-witness to that. One little acquaintance made a nest of yellow down and put it among the green oak leaves, making me think that the laws of protective coloration had no weight with her, but before the eggs were laid she had neatly covered the yellow with flakes of green lichen. I found her one day sitting in the sun with the top of her head as white as though she had been diving into the flour barrel. Here was one of the wonderful cases of 'mutual help' in nature. The flowers supply insects and honey to the hummingbirds, and they, in turn, as they fly from blossom to blossom probing the tubes with the long slender bills that have gradually come to fit the shape of the tubes, brush off the pollen of one blossom to carry it on to the next, so enabling the plants to perfect their flowers as they could not without help. It is said that, in proportion to their numbers, hummingbirds assist as much as insects in the work of cross-fertilization.

Though this little hummer that I was watching let me come within a few feet of her, when a lizard ran under her bush she craned her neck and looked over her shoulder at him with surprising interest. She doubtless recognized him as one of her egg-eating enemies, on whose account she put her nest at the tip of a twig too slender to serve as a ladder.

Another hummingbird who built across the

way was still more trustful — with people. I used to sit leaning against the trunk of her oak and watch the nest, which was near the tip of one of the long swinging branches that drooped over the trail. When the tiny worker was at home, a yard-stick would almost measure the distance between us. As she sat on the nest she sometimes turned her head to look down at the dog lying beside me, and often hovered over us on going away.

The nest was saddled on a twig and glued to a glossy dark green oak leaf. Like the other nest, it was made of a spongy yellow substance, probably down from the underside of sycamore leaves; and like it, also, the outside was coated with lichen and wound with cobweb. The bird was a rapid worker, buzzing in with her material and then buzzing off after more. Once I saw the cobweb hanging from her needle-like bill, and thought she probably had been tearing down the beautiful suspension bridges the spiders hang from tree to tree.

It was very interesting to see her work. She would light on the rim of the nest, or else drop directly into the bottom of the tiny cup, and place her material with the end of her long bill. It looked like trying to sew at arm's length. She had to draw back her head in order not to reach beyond the nest. How much more convenient it would have been if her bill had been jointed! It seemed better suited to probing flower tubes than making nests. But then, she made nests only in spring, while she fed from flowers all the year round, and so could afford to stretch her neck a trifle one month for the sake of having a good long fly spear during the other eleven. The peculiar feature of her work was her quivering motion in moulding. When her material was placed she moulded her nest like a potter, twirling around against the sides, sometimes pressing so hard she ruffled up the feathers of her breast. She shaped her cup as if it were a piece of clay. To round the outside, she would sit on the rim and lean over, smoothing the sides with her bill, often with the same peculiar tremulous motion. When working on the outside, at times she almost lost her balance, and fluttered to keep from falling. To turn around in the nest, she lifted herself by whirring her wings.

When she found a bit of her green lichen about to fall, she took the loose end in her bill and drew it over the edge of the nest, fastening it securely inside. She looked very wise and motherly as she sat there at work, preparing a home for her brood. After building rapidly she would take a short rest on a twig in the sun, while she plumed her feathers. She made nest-making seem very pleasant work.

One day, wanting to experiment, I put a handful of oak blossoms on the nest. They covered the cup and hung down over the sides. When the small builder came, she hovered over it a few seconds before making up her mind how it got there and what she had better do about it. Then she calmly lit on top of it! Part of it went off as she did so, but the rest she appropriated, fastening in the loose ends with the cobweb she had brought.

She often gave a little squeaky call when on the nest, as if talking to herself about her work. When going off for material she would dart away and then, as if it suddenly occurred to her that she did not know where she was going, would stop and stand perfectly still in the air, her vibrating wings sustaining her till she made up her mind, when she would shoot off at an angle. It seemed as if she would be worn out before night, but her eyes were bright and she looked vigorous enough to build half a dozen houses.

"There 's odds in folks," our great-grandmothers used to say; and there certainly is in bird folks; even in the ways of the same one at different times. Now this hummingbird was content to build right in front of my eyes, and the hummer down at the little lover's tree, with her first nest, was so indifferent to Billy and me that I took no pains to keep at a distance or disguise the fact that I was watching her. But when her nest was destroyed she suddenly grew old in the ways of the world, and apparently repented having trusted us. In any case, I got a lesson on being too prying. The first nest had not been down long before I found that a second one was being built only a few feet away — by the same bird? I imagined so. The nest was only just begun, and being especially interested to see how such buildings were started, I rode close up to watch the work. A roll of yellow sycamore down was wound around a twig, and the bottom of the nest — the floor — attached to the underside of this beam; with such a solid foundation, the walls could easily be supported.

The small builder came when Billy and I were there. She did not welcome us as old friends, but sat down on her floor and looked at us — and I never saw her there again. Worse than that, she took away her nest, presumably to put it down where she thought inquisitive reporters would not intrude. I was disappointed and grieved, having already planned — on the strength of the first experience — to have the mother hummer's picture taken when she was feeding her young on the nest.

At first I thought this suspicion reflected upon the good sense of hummingbirds, but after thinking it over concluded that it spoke better for hummingbirds than for Billy and me. If this were, as I supposed, the same bird who had to brood her young with Billy grazing at the end of her bill, and if she had been present at the unlucky moment when he got the oak branches tangled in the pommel of the saddle, although her branch was not among them, I can but admire her for moving when she found that the Philistines were again upon her, for her new house was hung at the tip of a branch that Billy might easily have swept in passing.

These nests had all been very low, only four or five feet above the ground; but one day I found young in one of the common treetop nests. I could see it through the branches. Two little heads stuck up above the edge like two small Jacks-in-boxes. Billy made such a noise under the oak when the bird was feeding the youngsters that I took him away where he could not disturb the family, and tied him to an oak covered with poison ivy, for he was especially fond of eating it, and the poison did not affect him.

Before the old hummer flew off, she picked up a tiny white feather that she found in the nest, and wound it around a twig. On her return, in the midst of her feeding, she darted down and set the feather flying; but as it got away from her, she caught it again. The performance was repeated the next time she came with food; but she did it all so solemnly I could not tell whether she were playing or trying to get rid of something that annoyed her.

She fed at the long intervals that are so trying to an observer, for if you are going to sit for hours with your eyes glued to a nest, it really is pleasant to have something happen once in a while! Though the mother bird did not go to the nest often, she sometimes flew by, and once the sound of her wings roused the young, and they called out to her as she passed. When they were awake, it was amusing to see the little midgets stick out their long, thread-like tongues, preen their pin-feathers, and stretch their wings over the nest.

One fine morning when I went to the oak I heard a faint squeak, and saw something fluttering up in the tree. When the mother came, she buzzed about as though not liking the look of things, for her children were out of the nest, and behold! — a horse and rider were under her tree. She tried to coax the unruly nestlings to follow her into the upper stories, but they would not go.

Although not ready to be led, one of the infants soon felt that it would be nice to go alone. When a bird first leaves the nest it goes about very gingerly, but this little fellow now began to feel his strength and the excitement of his freedom. He wiped his tongue on a branch, and then, to my astonishment, his wings began to whirl as if he were getting up steam, and presently they lifted him from his twig, and he went whirring off as softly as a hummingbird moth, among the oak sprays. His nerves were evidently on edge, for he looked around at the sound of falling leaves, started when Billy sneezed, and turned from side to side very apprehensively, in spite of his out-in-the-world, big-boy airs. He may have felt hampered by his unused wings, for, as he sat

The Swing Nest of the Hummer.
(From a Photograph.)

there waiting for his mother to come, he stroked them out with his bill to get them in better working order. That done, he leaned over, rounded his shoulders, and pecked at a leaf as if he were as much grown up as anybody.

Of all the beautiful hummingbirds' nests I saw in California, three are particularly noteworthy because of their positions. One cup was set down on what looked like an inverted saucer, in the form of a dark green oak leaf wound with cobweb. That was in the oak beside the ranch-house. Another one was on a branch of eucalyptus, set between two leaves like the knot in a bow of stiff ribbon. To my great satisfaction, the photographer was able to induce the bird to have a sitting while she brooded her eggs. The third nest I imagined belonged to the bird who took up her floor because Billy and I looked at her. If she were, her fate was certainly hard, for her eggs were taken by some one, boy or beast. Her nest was most skillfully supported. It was fastened like the seat of a swing between two twigs no larger than knitting-needles, at the end of a long drooping branch. It was a unique pleasure to see the tiny bird sit in her swing and be blown by the wind. Sometimes she went circling about as though riding in a merry-go-round; and at others the wind blew so hard her round boat rose and fell like a little ship at sea.

XIII.

IN THE SHADE OF THE OAKS.

THERE were half a dozen places in the valley, irrigated by the spring rains, where I was always sure of finding birds. Among them, on the west side, was the big sycamore, standing at the lower end of the valley; while above, in the northwest corner, was the mouth of Twin Oaks canyon where the migrants flocked in the brush around the large twin oak that overlooked the little old schoolhouse. On the east side was the Ughland canyon, at the mouth of which the little lover and his neighbors nested; while below it straggled the line of sycamores that followed the Ughland stream down through my ranch. But up at the head of the valley beyond the ranch-house was the most delightful place of all. There I was always sure of finding interesting nests to study.

Surrounded by a waste of chaparral, it was a little oasis of great blooming live-oaks, and in their shade I used often to spend the hot afternoon hours. In the spring the water that flowed down the hills at the head of the valley formed a fresh mountain stream that ran down the Oden canyon and so on through the centre of this grove, feeding the oaks and spreading out to enrich the valley below. In summer, like the rest of the canyon streams, only its dry sandy bed remained. Then, when the meadows were oppressively hot, my leafy garden was a shady bower to linger in. Its long drooping branches hung to the ground, dainty yellow warblers flitted about the golden tassels of the blossoming trees, and the air was full of the happy songs of mated birds.

The trail from the ranch-house to the oaks was a line through the low grass in which grew yellow fly flowers and orange poppies; and over them every spring, day after day, processions of migrating butterflies drifted slowly up the canyon. At the entrance of the garden was a sentinel oak whose dark green foliage contrasted well with the yellow flowers in the grass outside. It was the chosen hunting-ground of many birds. Its dead upper branches offered the bee-birds and woodpeckers an unobstructed view of passing insects, and gave the jays and flickers a chance to overlook the brush and take their bearings. The lower limbs offered perches where doves might come to rest, finches to chatter, and chewinks to sing; while its hanging boughs and elm-like feathered sides attracted wandering warblers and songful wrens.

The happy days spent among these beautiful

A SHADY BOWER

California oaks are now far in the past, but as I sit in my study in the East and dream back over those hours my mind is filled with memory pictures. Sauntering through this oaken gallery, each tree recalls some pleasant hour — the sight of a new bird, the sound of a new song, the prolonged delight of some cozy home that I watched till accepted as a friend, when the little family's fears and joys were my own.

That big double oak, spreading across the middle of the garden, was the haunted tree whose blue ghost drove away the pewees and gnatcatchers after they had begun to build; though the vireos and bush-tits braved it out, and the tiny hummer and gentle dove were not afraid to perch there. This was hummingbird lane — that small oak held the nest in which the two wee nestlings sat up like Jacks-in-the-box; these blue sage bushes growing in the sand were the ones the honey bees and hummers used to haunt, the hummers probing each lavender lip as they circled round the whorls; in front of this bush I saw a fairy dancer perform his airy minuet, — swing back and forth, and then sweep up in the air to dive whirring down with gorget puffed out and tail spread wide; and here, when watching a procession of ants, I discovered a tiny hummingbird building in a drooping branch that overhung the trail. That dead limb was the perch of a wood pewee, a silent grave bird with a sad call, who flew on when he was still only a lonely stranger. That oak top was made memorable by the sight of a flaming oriole, though he came on a cold foggy morning and answered my calls with a broken song and a half-hearted scold as he sat with his feathers ruffled up about him. Under the low spreading branches of that tree the chewinks used to scratch — I can hear the brown leaves rustle now — the branches were so low that, if the shy birds flew up to rest from their labors, they could quickly drop down and disappear in the brush.

On ahead, where the garden narrows to the trail between the walls of brush, when I was hidden behind a screen of branches, the timid white-crowned sparrows used to venture out, hopping along quietly or stopping to sing and pick up seeds on the path. Back a few steps was the tree where the bush-tits came to build their second nest after the roof of the first one fell in; the nest which hung on such a low limb that I watched it from the sand beneath, looking up through the branches at the blue sky, the canyon walls covered with sun-whitened bowlders, and the turkey buzzards circling over the mountains.

Just there, in that small open place between the trees, — how well I remember the afternoon, — I saw a new bird come out of the bushes; the green-tailed chewink he proved to be, on his way back to the Rocky Mountains. He was a beautiful stranger with a soft glossy coat touched off with yellowish green, while his high-bred gentle manners have made me remember him with affectionate interest all these years. Across the garden I heard my first song from that unique rhapsodist, the yellow-breasted chat. The same place marks another interesting experience. While I was sitting in the crotch of an oak a thrasher came out of the brush into an open space in front of me. Her feathers were disordered and apparently she had come from her nest. She walked with wings tight at her sides and her tail up at an angle well out of the way of the rustling leaves; altogether a neat alert figure that contrasted sharply with the lazy brown chippie which appeared just then in characteristic negligée, its wings hanging and tail dragging on the ground. The thrashers of Twin Oaks have bills that are curved like a sickle, and this bird used her tool most skillfully. Instead of scratching up the leaves and earth with her feet as chewinks and sparrows do, the thrasher used her bill almost exclusively. First she cleared a space by scraping the leaves away, moving her bill through them rapidly from side

Green-tailed Chewink.
(One half natural size.)

to side. Then she made two holes in the ground, probing deep with her long bill. After taking what she could get from the second hole, she went back to the first again, as if to see if anything had come to the surface there. Then she lay down on the sand to sun herself and acted as though going to take a sun bath, when suddenly she discovered me and fled.

When watching the bird at work I got a pretty picture in the round disk of my opera-glass. The glass was focused on the digging thrasher, but a goldfinch came into the picture and pulled at some stems for its nest and a cottontail ran rapidly across from rim to rim. I lifted the glass to follow him and saw him go trotting down the path between the bushes.

The thrasher's curved bill gives a most ludicrous look to the bird when singing. He looks as if he were trying to turn himself inside out. I once saw an adult thrasher tease its mate for food, and wondered how it would be possible for one curved bill to feed another curved bill; but a few days later I came on a family of young, and discovered for myself that *they* have straight bills; a most curious and interesting instance of adaptation.

At the head of the garden stands a tree that always reminds me of the horses I rode in California. I watched my first bush-tit's nest under it, with Canello grazing near; and five years later watched another bush-tit's nest there, sitting in the crotch of the oak with Mountain Billy looking over my shoulder. Although Billy was, in his prime, a bucking mustang, he became more of a petted companion than Canello had been; and when we were out alone together, we were a great deal of company for each other. As soon as I dismounted he would put his head down to have me slip the reins off over his ears, so that he could graze by himself. Sometimes, when he stood behind me he rested his bridle on my sun-hat, and once went so far as to take a bite out of the brim — in consideration of its being straw. If I were sitting on the ground and he was grazing near, he would at times walk up and gravely raise his face to look into mine. When he got tired, he would rub up against my arm and yawn, looking down at me with a friendly smile in his eyes.

Birding was rather dull for Billy — when there was neither grass nor poison ivy at hand, but he had one never-failing source of enjoyment — rolling. He tried it in the sand under the oak, one day, with the saddle on. Before I knew what he was about he was down on his knees, sitting still, with a comical, helpless look in his eyes, as if quite at a loss to know what to do next, having become conscious of the saddle. When I had gotten him on his feet and finished lecturing him I uncinched the saddle, laid it one side on the ground, took hold of the end of the long bridle, and told him to roll. A droll abstracted look came into his eyes, he dropped on his knees and, with a sudden convulsion, threw his heels into the air and rolled back and forth, rubbing his backbone vigorously on the sand. After that, the first thing every morning when we got to the oaks, I unsaddled him and let him roll, and then he would stand with bare back keeping cool in the shade of the trees.

One morning as we stood under the bush-tit's tree, I discovered a pair of turtle doves looking out at me from the leaves of the small oak opposite, craning their necks and moving their heads uneasily. One of them seemed to be shaping a nest of twigs. I drew Billy around between us, so that my staring would seem less pointed, and when one of the pair flew to the ground to spy at me, hurriedly looked the other way to remove his anxiety. His mate soon joined him, and the two doves walked away together, fixed their feathers in the sun, stretched their wings, and lazily picked at the ground. When one whirred back to the nest, the other soon followed. The gentle lovers put their bills together, while, unnoticed, I stood behind Billy, looking on and thinking that it was little wonder such birds should rise from the ground with a musical whirr.

Billy's oak was the last of the high trees in the garden. Above it was a grassy space where bright wild flowers bloomed, and pretty cottontail rabbits often went ambling over the soft turf. On one side of the opening was a low stocky oak, full of balls of mistletoe, and on the other a great blossoming bush buzzing with hummingbirds. The mistletoe had begun to sap the little oak, and on one of its dead twigs a hummingbird had taken to perching. I wondered if he were the idle mate of one of my small garden builders, but he sat and sunned himself as if his conscience were quite clear.

My first experience with gnatcatchers had been here. I suspected a nest, and the ranchman's daughter went with me to hunt through the brush. She cautioned me to look out for rattlesnakes, but the brush was so dense and the ground so covered with crooked snake-like sticks that it was not an easy matter to tell what you were stepping on. Then, the poison oak was so thick that I felt like holding up my hands to avoid it. We pushed our way through the dense chaparral, and my fearless companion got down on her hands and knees to look through the tangle for the nest. It was hard disagreeable work, even if one did not object to snakes, and we were soon so tired that we were ready to sit down and let the birds show us to their house. We might have saved ourselves all the trouble if we had done this to begin with, for it was only a few moments before the little pair went to the mistletoe oak, out in plain sight and within easy reach — how they would have laughed in their sleeves had they known what we were hunting for back in the brush! The nest was about the size of a chilicothe pod, and so covered with lichen that it looked just like a knot on the tree.

Around the blossoming bush the air fairly vibrated with hummers, darting up into the sky, shooting down and chasing each other pell mell — sometimes almost into my face. As I sat by the bush one day, a handsome male went around with upraised throat, poking his bill up the red fuchsia-like tubes. Another one was flying around inside the bush, and I edged nearer to see. The sun shone in, whitening the twigs, and as the bird whirred about with a soft burring sound, I caught gleams of red, gold, and green from his gorget, and could see the tiny bird rest his wee feet on a twig to reach up to a blossom. Then he hummed what sounded more like a love song than anything I had ever heard from a hummingbird. He seemed so much more like a real bird than any of his brothers that I felt attracted to him.

One morning a little German girl, in a red pinafore, and with hair flying, came riding down the sand stream toward my bush. Her colt reared and pranced, but she sat as firmly as if she had been a small centaur. It was a holiday, and she was staking out her horses to graze, making gala-day work of it. She had one horse down by the little oak already, and springing off the one she had brought, changed about, jumped as lightly as a bird upon the other's back and raced home. Soon she came galloping back again, and so she went and came until tired out, for pure fun on her free holiday.

In looking over the bright memory pictures of my beautiful oak garden, there is one to which I always return. The spreading trunks of a great five-stemmed tree on one side of the grove made a dark oaken couch, screened by the leafy willow-like branches that hung to the ground. Here — after looking to see that there were no rattlesnakes coiled in the dead leaves — I spent many a dreamy hour, reclining idly as I listened to the free songs of the birds that could not see me behind my curtain. It was interesting to note the way certain sounds predominated; certain songs would absorb one's attention, and then pass and be replaced by others. At one time a jay's scream would jar on the ear and drown all other voices; when that had passed, the chewinks would fly up from the leaves and sing and answer each other till the air was quivering with their trills. Then came the thrashers, with their loud rollicking songs; and when they had pitched down into the brush, out rang the clear bell-like tones of the wren-tit, filling the air with sound. Afterwards the impatient whipped-out notes of the chaparral vireo were followed by the soft cooing of doves; and then, as the wind stirred the trees and sent the loosened oak blossoms drifting to the ground, from high out of an oak top came a most exquisite song. At the first note of this grosbeak all other songs were forgotten — they were noise and chatter — this was pure music. It was like passing from the cries of the street into the hall of a symphony concert. The black-headed grosbeak has not the spirituality of the hermit thrush, and his ordinary song is not so remarkable, but his love song excels that of any bird I have ever heard in finish, rich melody, and music. As I listened, my surroundings harmonized so perfectly with the wonderful song echoing through the great trees that the old oak garden seemed an enchanted bower. The drooping branches were a leafy lattice through which the afternoon sun filtered, steeping the oaks in thick still sunshine. Last year's leaves drifted slowly to the ground, while the bees droned about the yellow tassels of the blooming trees. As a violinist, lingering to perfect a note, draws his bow again and again over the strings, so this rapt musician dwelt tenderly on his highest notes, trolling them over till each was more exquisite and tender than the last, and the ear was charmed with his love song — a song of ideal love fit to be dreamed of in this stately green oak garden filled with golden sunlight.

XIV.

A MYSTERIOUS TRAGEDY.

On a peg just inside the door of the ranchman's old wine shed hung one of the horses' unused nosebags. A lad on the place told me that a wren had a nest in it, and added that he had seen a fight between the wren and a pair of linnets who seemed to be trying to steal her material.

The first time I went to the wine shed both wrens and linnets were there, but nothing happened and I forgot about the original quarrel. By peering through a crack in the boarding I could look down on the wren in the nosebag inside. I could see her dark eyes, the white line over them, and her black barred tail. She was Vigor's wren. She got so tame that she would not stir when the creaking door was opened close by her, or when people were talking in the shed; and I used to go often to see how her affairs were progressing.

All her eggs hatched in time, and the small birds, from being at first all eyeball, soon got to be all bill. When I opened the bag to look at them, the light woke them up and they opened their mouths, showing chasms of yellow throat.

The mother bird fed them several times when I was watching only a few feet away. She would come ambling along in the pretty wren fashion, with her tail over her back; creeping down the side of a lath, running behind a rafter, scolding as though to make conversation, and then winding down to the nest through a crack. One day she hesitated, and waited to spy at me, since I had thought it polite to stare at her! When satisfied, she hopped along from beam to beam, her bright eyes still upon me. Then her mate joined her. He had been suspicious of me at our first meeting, but apparently had changed his mind, for, seeing his spouse hesitate, he glanced at me unconcernedly, as much as to say, "Is she all you 're waiting for?" and flew out, leaving her to my tender mercies. She hopped meekly into the bag after that rebuke, but stretched up to peer at me once more before settling down inside.

One day when I looked in to see how wren matters were progressing, to my amazement and horror, instead of my wren's nest I found another, high in the mouth of the bag with one fresh egg in it! The egg was a linnet's, and the nest had been built right on top of the wren's. Such a stench came from the bag that I took out the upper nest and found the four little wrens dead in their crib.

I had become very fond of the winsome mother bird, and so much interested in her brood that

The Nosebag Nest.

(Vigors's Wren.)

this horrid discovery came like a tragedy in the family of a friend.

And what did it all mean? Unless the old wrens had been dead, could the linnets have gotten possession? The wrens were usually able to hold their own in a discussion. If the nestlings had been alive, would the linnets — would any bird — have built upon them, deliberately burying them alive? It seemed too diabolical. On the other hand, what could have killed the little wrens and left them in the nest? If they had been dead when the linnets came to build, how could the birds have chosen such a sepulchre for a building site?

Grieving over my little friends, I cleaned out the nosebag and hung it up on its peg. Three weeks later I discovered, to my great perplexity, that a pair of wrens had built in the bottom of the bag and had one egg in the nest. Now, was this the same pair of birds that had built there before, and if so, what did it all mean?

XV.

HOW I HELPED BUILD A NEST.

THEY picked out their crack in the oak and began to build without any advice from me, winning little gray-crested titmice that they were. Their oak was right behind the ranch-house barn; I found it by hearing the bird sing there. The little fellow, warmed by his song, flitted up the tree a branch higher after each repetition of his loud cheery *tu-whit′*, *tu-whit′*, *tu-whit′*, *tu-whit′*. Meanwhile his pretty mate, with bits of stick in her bill, walked down a crack in the oak trunk.

Thinking she had gone, I went to examine the place. I poked about with a twig but could n't find the nest till, down in the bottom of the crack, I spied a little gray head and a pair of bright eyes looking up at me. The bird started forward as if to dart out, but changed her mind and stayed in while I took a hasty look and fled, more frightened than she by the intrusion.

The titmice had been flying back and forth from the hen-yard with chicken's feathers, and it seemed such slow work for them I thought I would help them. So the next day, when the pair were away, I stuffed a few white feathers into the mouth of the nest and withdrew under the shadow of the barn to watch through my glass without being observed. Then my conscience began to trouble me. What if this inter-

The Plain Titmouse in her Doorway.

ference should drive the gentle bird to desert her nest?

When I heard the familiar chickadee call — the titmouse often chirrups like his cousin — it made me quake guiltily. What would the birds do? The gray pair came flying in with crests raised, and my small friend hopped down to her doorway. She gave a start of surprise at sight of the feathers, but after a moment's hesitation went bravely in! While she was inside, her mate waited in the tree, singing for her; and when she came out, he flew away with her. Then I crept up to the oak, and to my delight found that all the feathers had disappeared. She evidently believed in taking what the gods provide. In fact, she seemed only to wish that they would provide more, for, after taking a second supply from me, she stood in the vestibule, cocked her crested head, and looked about as if expecting to see new treasures.

She had common-sense enough to take what she found at hand, but if she had not been such a plucky little builder she would have been scared away by the strange sights that afterwards met her at her nest. Once when she came, feathers were sticking in the bark all around the crack. She hesitated — the rush of her flight probably fanned the air so the white plumes waved in her face — she hesitated and looked around timidly before getting courage to go in; and on leaving the nest flew away in nervous haste; but she was soon back again, and ready to take the feathers down inside the oak. She caught hold of the tip of one that was wedged into a crack, and tugged and tugged till I was afraid she would get discouraged and go off without it. She got it, however, and drew it in backwards. Then she attacked another feather, but finding that it came harder than the first, let go her hold and took an easier one. She was not to be daunted, though, and after stowing away the loose one came back for the tight one again, and persevered till she bent it in several places, besides breaking off the tip.

When she had flown off, I jumped up, ran to the oak, and stuffed the doorway full of feathers. Before I had finished, the family sentinel caught me — I had been in too much of a hurry and he had heard me walking over the cornstalks. He eyed me suspiciously and gave vent to his disapproval, but I addressed him in such friendly terms that he soon flew off and talked to his mate reassuringly, as if he had decided that it was all right after all. After their conversation she came back and made the best of her way right down through the feather-bed! I went away delighted with her perseverance, and charmed by her confidence and pretty performances.

The next day I heard the titmouse singing in an elder by the kitchen, and went out to see how the birds acted when gathering their own material. The songster was idly hunting through the branches, singing, while his mate — busy little housewife — was hard at work getting her building stuff. She had something in her beak when I caught sight of her, but in an instant was down on the ground after another bit. Then she flew up in the tree looking among the leaves; in passing she swung a moment on a strap hanging from a branch; then flew down among the weeds, back up in the tree again; and so back and forth, over and over, her bill getting fuller and fuller.

I was glad to save her work, and interested to see how far she would accept my help. Once when I blocked the entrance with feathers and horsehair she stopped, and, though her bill was full, picked up the packet and flew out on a branch with it. Was she going to throw away my present? For a moment my faith in her was shaken. Perhaps her mate had been warning her to beware of me. She did drop the mat of horsehair — what did such a dainty Quaker lady as she want of horsehair? — but she kept tight hold of one of the feathers, although it was almost as big as she was, and flew back quickly to the nest with it.

This performance proved one point. She would not take everything that was brought to her. She preferred to hunt for her own materials rather than use what she did not like. Now the question was, what did she like?

My next experiment was with some lamp wick to which I had tied bits of cotton. The titmouse took the cotton and would have taken the wicking, I think, if it had not been fastened in too tight for her. After that I tried tying bits of cotton to strings, and letting them dangle before the mouth of the nest. Though I moved up to within twenty feet of the nest, she paid no attention to me but hurried in. She liked the cotton so well she stopped in her hallway, reached up to pull at the white bundles, and tweaked and tugged till, finally, she backed triumphantly down the hole with one.

Her mate, less familiar with my experiments, started to go to the nest after her, but the sight of the cotton scared him so he fled ignominiously back into the treetop. He stayed there singing till she came out, when he flew up to her with a dainty he had discovered — at least the two put their bills together; perhaps it was just a caress, for they were a tender, gentle little pair.

Having proved that my bird liked feathers and cotton, I wanted to see what she thought of straws. Apparently she did not think much of them. She looked very much dashed when she came home and found the yellow sticks protruding from the nest hole. She hesitated, turned her head over, flew to a twig on one side of the oak and then back to one on the other side. Finally she mustered courage, and with her crest flattened as if she did not like it, darted down into the hole. When she flew out, however, she went right to her mate, and forgetting all her troubles at sight of him, fluttered her wings and lisped like a young bird as she put up her bill to have him feed her.

Perhaps it was unkind to bother the poor bird

any more, but I meant her no harm and the fever for experiment possessed my blood. I tied some of the straws to a piece of wicking and baited it with feathers, thinking that perhaps she would take the straws for the sake of the feathers and wicking. I also stuffed the hole with horsehair. She did pull at the feather end of the line; I saw the straw jerk, and, when she had left, found a round hole the brave little bird had made right through the middle of the mat of horsehair I had stopped the nest with.

Straws and horsehair the titmouse evidently classed together. They were not on her list of building materials. On reflection she decided that the horsehair would make a good hall carpet, so left it in the vestibule, though she would have none of it down in her nest; but she calmly threw my straws down on the ground at the foot of the oak.

I don't know what experiments I might have been tempted to try next had I not suddenly found myself dismissed — the house was complete. My pretty Quaker lady sat in the shade of the oak leaves with crest raised and the flickering sunlight flecking her gray breast. She pecked softly at one of the white feathers that blew up against her as she listened to the song of her mate; and then flew away to him without once going to the nest. Evidently her work was done, and she was waiting till it should be time to begin brooding.

Ten days later I saw her mate come with his bill full of worms and lean down by the hole to call her. She answered with a sweet pleading twitter, and reached up to be fed. When he had gone, perhaps she thought she would like a second bite. At any rate, she hopped out in the doorway and flew off to another tree, calling out *tsché-de-de* so sweetly he would surely have come back to her had he been within hearing.

A few days later I saw him feed her at the nest five or six times in half an hour. He would come to the next oak, light and call to her, when she would answer from inside the tree trunk and he would go to her. I was near enough to see her pretty gray head and black eyes coming up out of the crack in the oak. Sometimes when he had fed her he would call out and she would answer as if saying good-by from down in the nest. One morning I found the devoted little mate bringing her breakfast to her at half past six.

Nearly a month later they were feeding their young. The winsome mother bird, who had looked so tired and nest-worn the last time I saw her, was now as plump and happy as her spouse. When I thought the pair were away, I went to try to get sight of the nestlings down the hole. The old birds appeared as soon as I set foot by the oak and took upon themselves to scold me. They chattered softly in a way they had never done before. They quickly got used to me again, however, and fed the little ones without hesitation right before me, knowing full well that a person who had helped them build their nest would never harm their little brood; and it was a disappointment when I had to go away and leave the winning family.

XVI.

IN OUR NEIGHBOR'S DOOR-YARD.

THE little German girl with the scarlet pinafore was a near neighbor, living at the head of the valley in a cottage surrounded by great live-oaks. These trees were alive with birds. Bush-tits flew back and forth, busily hanging their gray pockets among the leafy folds of the drooping branches; blue jays flew through, squawking on their way to the brush; goldfinches, building in the orchard, lisped sweetly as they rested in the oaks; and a handsome oriole who was building in the grove flew overhead so slowly he seemed to be retarded by the fullness of his own sweet song. But I had become so fond of the gentle gray titmouse whose nest I had helped to build, that of all the bird songs in the trees, its cheery *tu-whit′*, *tu-whit′*, *tu-whit′* was most enticing to me. How delightful it would be to watch another pair of the winning workers! I did see one of the birds enter a hollow branch, one day, and not long after saw it go down a hole in an oak trunk; but never saw it afterwards in either place. Back and forth I followed that elusive voice, hoping to discover the nest, but I suspect the bird was only prospecting, and had not even begun to work.

The little German Gretchen became interested in the search for the titmouse's nest, and told me that a gray bird had built in an oak in front of her house. I rode right over to see it, but found the gray bird a female Mexican bluebird, whose brilliant ultramarine mate sat on the fence of the vegetable garden in plain sight. The children kept better watch of the nest after that, and a few days later, when in my attic study, I heard the tramp of a horse, and, looking out, found my little friend under the window, come to tell me that the eggs had hatched. When her older sister came for the washing I asked her if she had seen the old birds go to the nest, and she said, "Yes; one was blue and the other gray."

When I rode up again, the young had grown so that from the saddle I could look down the hole and see their big mouths and bristling pinfeathers. The mother bird was about the tree, and her soft dull coloring toned in well with the gray bark. The bluebirds had a double front door, and went in one side to come out the other. I saw both of them feed the young, the male flying into the hole straight from the fence post.

It seemed such hard work finding worms out in the hot sun that I wondered if birds' eyes ever ached from the intentness of their search, and if there were near-sighted birds. Perhaps the intervals of feeding depend on the worm supply rather than the dietary principles of the parents.

Gretchen's mother was bending over her wash-tubs out under the oaks, and I called her attention to the pretty birds brooding in her door-yard, telling her that they were good friends of hers, eating up the worms that destroyed her flowers and vegetables. "So?" she asked, but seemed ready to let the subject drop there, and hurried back to her work. A poor widow with a large family of children and a ranch to look after can find little time, even in beautiful California, to enjoy what Nature places in her door-yard.

Three weeks later Gretchen came riding down to tell me that there were eggs in the tree again. The bluebird bid fair to be as hardworked as the widow, at that rate, I thought, when I went up to look at them. The children showed me the nest of a goldfinch, near the ground, in one of the little orange-trees in front of the house. They also pointed out linnets' nests in the vines by the door, and the oldest child said eagerly, "When we came home from school there was a humming-bird in the window, and we caught it," adding, "I think it must have been a father humming-bird." "Why?" I asked, "was it pretty?" "Yes, it just shined," she exclaimed enthusiastically.

When the family were at home, their puppy would bark at us furiously, and follow us about suspiciously, but when he had been left on the ranch alone he was glad of our society. Then when I watched the bluebirds, he came and curled down by my side, becoming so friendly that he actually grew jealous of Billy, and turned to have me caress him each time that the little horse walked up to have the flies brushed off his nose, or having pulled up a bunch of grass by the roots, brought it for me to hold so that he could eat it without getting the dirt in his mouth.

Going home one day, Billy came upon a gopher snake. Now Canello had been brought up in a rattlesnake country, and was always on his guard, but Billy was 'raised' in the mountains, where snakes are scarce, and did not seem to know what they were. He had given me a good deal of anxiety by this indifference — he had stepped over a big one once without seeing any need for haste — and I had been expecting that he would get bitten. Here, then, was my chance to give him a scare. The gopher snake was harmless; perhaps, if I could get him so close to it that he would see it wriggle away from under his feet, he might be less indifferent to rattlers.

The gopher snake was three or four feet long, and lay as straight as a stick across our path. As I urged Billy up beside it, he actually stepped on the tip of its tail. The poor snake writhed a little, but gave no other sign of pain; its rôle was to remain a stick. And Billy certainly acted as if it were. I threw the reins on his neck, thinking that if he put his head down to graze he might make a discovery. Then a horrid thought came to me. The people said the rattlers sometimes lost their rattles. In a general way, rattlers and gopher snakes look alike; what if this were a rattlesnake, and at my bidding my little horse should be struck! But no. There was no mistaking the long tapering body of the gopher, and it lacked the wide flat head of the rattler. But I might have spared myself my fears. Billy would not even put his head down, and when I tried to force him upon the snake he quietly turned aside. To make the snake move, I threw a stick at it, but it was as obstinate as Billy himself. Then I slipped to the ground, and picking up a long pole gave it a gingerly little poke. Still motionless! I tried another plan, taking Billy away a few yards. Then at last the snake slowly pulled itself along. But the moment we came back it turned into a stick again, and Billy relapsed into indifference. It was no use. I could do nothing with either of them. I would see the snake go off, anyway, I thought, so withdrew and waited till it felt reassured, when it started. Its silken skin shone as it wormed silently through the grass and disappeared down a hole without a sound, and I reflected that it might also come *up* without a sound, very likely beside me as I sat on the dead leaves!

XVII.

WHICH WAS THE MOTHER BIRD?

THE second time I went to California the little whitewashed adobe opposite my ranch was still standing, but an acacia-tree had grown over the well where the black phœbe had nested, and the shaft was so overrun with bushes and vines that it was hard to find a trace of it. Drawn by pleasant memories, I rode in one morning, sure of finding something interesting about the old place.

I had not waited long before the chip of a young bird came from the vines over the well. It proved a callow nestling, with no tail, and little to mark its parentage. Presently a brown long-tailed wren-tit came with food in its bill and peered down through the leaves at it; and then a California towhee came and sat around till satisfied as to whose child was crying. A moment later a lazuli bunting flew over with food in her bill, and I at once bethought me of the lazuli-like markings, the brownish wing bars and the sharp cry of "quit," which none but a lazuli could give. That surely was my bird.

But if so, what did this interest on the part of the wren-tit mean? She hopped about the nestling with tail up and crest raised, chattering to it in low mysterious tones; and when I suspected her of giving her worm to it, suddenly turned her head and looked away with a suspiciously non-commital air. The lazuli, however, sat indifferently on a branch and plumed her feathers, though when she did fly down toward the young one, the wren-tit gave way. But even then the lazuli did not feed the small bird. When she had gone, the wren-tit came back. She spoke low to the nestling, and drew it down into the thick part of the tangle where I could not see them, though there was a hint of tiny quivering wings, and I was morally certain that the old bird was feeding it, especially when she flew up in sight with the smart air of having outwitted me.

I was getting more and more bewildered. What did it all mean? Were there two families of young down in the tangle? If not, why were two old birds feeding one little one, and to which mother did the child belong? The wisdom of Solomon was needed to solve the riddle.

The wren-tit simply devoted herself to the little bird, going and coming for it constantly; while the lazuli, ordinarily the most nervous noisy bird when her young are disturbed, sat around silently, or flew away without remark. I became so impressed by the wren-tit side of the case that I quite forgot the lazuli note and markings.

Just as I thought I had come to a decision in the case, a male lazuli flew in, lighting atilt of an acacia stalk opposite the wren-tit. But when he saw me he craned his neck and flew off in a hurry — no father, surely, scared away at the first glimpse of me! However, I was not clear in my mind, and sat down to puzzle the matter out.

At this juncture Madame Lazuli came with food; the young bird turned toward her for it, and behold! she took to her wings with all she had brought. I had hardly time to congratulate myself on this new piece of testimony, when back came the lazuli with her bill full!

In my perplexity I moved so near the little one that, without meaning to, I forced the old birds to show their true colors. The situation was too dangerous to admit of further subterfuge. Both Madame Lazuli and her handsome blue mate — whom I discovered at a safe distance up on a high branch out of reach — flew down and dashed about, twitching their tails from side to side as they cried "quit," in nervous tones; altogether acting so much like anxious parents that I had to relinquish my theory that the little bird belonged to the wren-tit. Like the mother whom Solomon judged, she forgot all else when real danger threatened the child. Having come to my decision from circumstantial evidence, I remembered with a start that I had known it all the time, from the wing-bars and the call note! Nevertheless, my riddle was only half solved, for how about the wren-tit?

A young bird called from the sycamore at the corner of the adobe, and when both old birds flew over to it, I thought I'd better follow. I got there just in time to see a little bird light in the elbow of a limb, totter as if going to fall, and save itself by snuggling up in the elbow, where it sat in the sun looking very cozy and comfortable — winning little tot. The mother lazuli started to come to it, but seeing me flew away to another branch, where, well screened, she stretched up on her toes to look at me over the top of a big sycamore leaf. Though the fledgling called, the mother left without going to it.

The wren-tit had stayed behind at the well; but while the lazuli was gone, who should come flying in but the foster mother! I was astonished. Moreover, the instant the youngster set eyes on her, it started up and flew to her — actually flew into her in its hurry. She admonished it gently, in a soft chattering voice, for she could not scold it.

When the lazuli came back with food, it was only to see her little bird flying off to the other side of the tree after the wren-tit! I thought she seemed bewildered, but she followed in their wake — we all followed. Here came a closer test. Both lazuli and wren-tit stood before the small bird. Which would it go to? The lazuli kept silent, but the wren-tit called softly and the little one raised its wings and flew toward her, leaving its mother behind.

I watched and waited, but the wren-tit did not give over her kind offices, and the last I saw of the birds, on riding away, the three were flying in procession across the brush, the lazuli following its mother and the wren-tit bringing up the rear.

I went home very much puzzled. Was the wren-tit a lonely mother bird who had lost her own little ones, or was she merely an old maid with a warm spot in her heart for other peoples' little folks?

XVIII.

A RARE BIRD.

WE may say that we care naught for the world and its ways, but most of us are more or less tricked by the high-sounding titles of the mighty. Even plain-thinking observers come under the same curse of Adam, and, like the snobs who turn scornfully from Mr. Jones to hang upon the words of Lord Higginbottom, will pass by a plain *brown chippie* to study with enthusiasm the ways of a *phainopepla!* Sometimes, however, in ornithology as in the world, a name does cover more than its letters, and we are duped into making some interesting discoveries as well as learning some of the important lessons in life. In the case of the phainopepla, no hopes that could be raised by his cognomen would equal the rare pleasure afforded by a study of his unusual ways.

On my first visit to Twin Oaks I caught but brief glimpses of this distinguished bird. Sometimes for a moment he lit on a bare limb and I had a chance to admire his high black crest and glossy blue-black coat, which with one more touch of color would become iridescent. He was so slenderly formed, and his shining coat was so

THE PHAINOPEPLAS ON THE PEPPER-TREE

smooth and trim, he made me think of a bird of glass perched on a tree. But while I gazed at him he would launch into the air and wing his way high over the valley to the hillsides beyond, leaving me to marvel at the white disks on his wings, hidden when perching, but in air making him suggest a black ship with white sails.

His appearance was so elegant and his ways so unusual that I went back East regretting I had not given more time to a bird who was so individual, and resolved that if I ever returned to California my first pleasure should be to study him. When the time finally came, an ornithologist friend who knew my plans wrote, exclaiming, "Do study the phainopeplas!" and added that she felt like making a journey to California to see that one bird.

From the middle of March till the middle of May I watched and waited for the phainopeplas. There had been only a few of the birds before, and I began to fear they had left the valley. When despairing of them, suddenly one day I saw a black speck cross over to the hills. I wanted to drop my work and follow, but went on with my rounds, and one bright morning on my way home after a discouraging hunt for nests, a pair of phainopeplas flew up right before my eyes almost within sight of the house. I dropped down behind a bush, and in a moment more the birds flew to a little oak by the road — a tree I had been sitting under that very morning! The female seated herself on top of the oak, watching me with raised crest, while her mate disappeared in a dark mat of leaves, probably mistletoe, where he stayed so long that the possibility of a nest waxed to a probability, and I made a rapid but ecstatic ascent to the observer's seventh heaven. A phainopepla's nest right on my own doorsill! I could hardly restrain my impatience, and was tempted to shoo the birds away so I could go to the nest; when suddenly they opened their wings and, crossing the valley, disappeared up a side canyon! Pulling myself together and reflecting that I might have known better than to imagine there would be a nest so near home, I took up my camp-stool and trudged back to the house.

After that came a number of tantalizing hints. When watching the third gnatcatcher's nest I had seen a pair of phainopeplas flying suggestively back and forth from the brush to the various oaks, and thought the handsome lover fed his mate as his relative the gentle high-bred waxwing does. Surely the wooing of these beautiful birds should be carried on with no less fine feeling, courtesy, and tenderness; and so it seems to be. The black knight flew low over my head slowly, as if inspecting me, and then came again with his lady, as if having said, "Dear one, I would consult you upon this impending danger."

After that, something really delightful came

about. Day by day, on riding back to our ranch-house, I found phainopeplas there eating the berries of the pepper-trees in our front yard. Before long the birds began coming early in the morning; their voices were the first sounds we heard on awakening and almost the last at night, and soon we realized the delightful fact that our trees had become the feeding ground for all the phainopeplas of the valley. Altogether there were five or six pairs. It was a pretty sight to see the black satiny birds perched on one of the delicate sprays of the willowy pepper-trees, hanging over the grape-like clusters, to pluck the small pink berries. The birds soon grew very friendly, and, though they gave a cry of warning when the cats appeared, became so tame they would answer my calls and let me watch them from the piazza steps, not a rod away.

When they first began to linger about the house we thought they were building near, and when one flew into an oak across the road, almost gave me palpitation of the heart by the suggestion. But no nest was there, and when the bird flew away it rose obliquely into the air perhaps a hundred feet, and then flew on evenly straight across to the small oaks on the farther side of a patch of brush that remained in the centre of the valley, known to the ranchmen as the 'Island.' The flight looked so premeditated that the first thing the next morning, although the phainopeplas were at the peppers, I rode on ahead to wait for them at their nest. We had not been there long before hearing the familiar warning call. Turning Billy in the direction of the sound, I threw his reins on his neck to induce him to graze along the way and give our presence a more casual air, while I looked up indifferently as if to survey the landscape. To my delight the phainopepla did not seem greatly alarmed, and, throwing off the assumed indifference that always makes an observer feel like a wretched hypocrite, I called and whistled to him as I had done at the house, to let him know that it was a familiar friend and he had nothing to fear. The beautiful bird started toward me, but on second thought retreated. I turned my back, but, to my chagrin, after giving a few low warning calls, my bird vanished. Alas, for the generations of murderers that have made birds distrust their best friends — that make honest observers tremble for what may befall the birds if they put trust in but one of the human species!

It was plain that if I would get a study of these rare birds I must make a business of it. Slipping from the saddle, I sat down behind a bush and waited. When the bird came back and found the place apparently deserted, to my relief he seated himself on a twig and sang away as if nothing had disturbed his serenity of spirit. But presently the warning call sounded again. This

THE PHAINOPEPLA'S NEST IN THE OAK BRUSH ISLAND

time it was for a schoolgirl who had staked out her horse on the edge of the island and was crossing over to the schoolhouse. A few moments later the bell rang out so loudly that Billy stepped around his oak with animation, but the phainopeplas were used to it and showed no uneasiness.

Before long a flash of white announced a second bird, and then, after a long interval in which nothing happened, the male pitched into a bush with beak bristling with building material! My delight knew no bounds. Instead of nesting in the top of an oak in a remote canyon, as I had been assured the shy birds would do, here they were building in a low oak not more than an eighth of a mile from the house, and in plain sight. Moreover, they were birds who knew me at home, and so would really be much less afraid than strangers, whatever airs they assumed. In the photograph, the bare twigs of the perch tree show above the line of the horizon; the nest tree is the low oak beside it on the right. One thing puzzled me from the outset. While the male worked on the nest, the female sat on the outside circle of brush as if having nothing to do, in spite of the fact that her gray dress toned in so well with the brush that she was quite inconspicuous, while his shining black coat made him a clear mark from a distance. What did it mean? I invented all sorts of fancies to account for it.

Had she been to the pepper-trees so much less than he that she was over-troubled by my presence, and therefore the gallant black knight who sang to her so sweetly and was so tender of her, seeing her fears, took the work upon himself? Perchance he had said, "If you are timid, my love, I will build for you while she is by, for I would not have you come near if it would disquiet you."

In any event, he built away quite unconcernedly not three rods from where I sat on the ground staring at him. He would fly to the earth for material, but return to the nest from above, pitching down to it as if having nothing to hide. Once, when resting, he perched on the tree, and I talked to him quite freely. That noon the phainopeplas were at the house before me, and I went out to talk to them while they lunched to let them know it was only I who had visited their nest, so they would have new confidence on the morrow.

But on the morrow they flew to another part of the island, and when we followed, although I hitched Billy farther away from the nest tree and sat quietly behind a brush screen, they did not come back. A brown chippie plumed his feathers unrebuked in their oak, making the place seem more deserted than before. A lizard ran out from the grape cuttings at my feet, and a little black and white mephitis cantered along over the ground with his back arched and his head down. He nosed around under the bushes, showing the white V on his back, exactly like that of our eastern species. As I rode home, five turkey buzzards were flying low over the edge of the island, and one vulture rose from a meal of one of the little black and white animal's relatives, but I saw nothing more of my birds that day.

The next day the phainopeplas came again to the pepper-trees and ate their fill while I sat on the steps watching. The male was quite unconcerned, but when his mate flew near me, he called out sharply; he could risk his own life, but not that of his love. Again the pair flew back to the high oaks on the far side of the island. All my hopes of the first low inaccessible nest vanished. I had driven the birds away. My intrusiveness had made me lose the best chance of the whole nesting season. But I would try to follow them. It did not seem necessary to take Billy. There were only a few trees on that side of the island, and it would be a simple matter to locate the birds. I would walk over, find in which tree they were building, and spend the morning with them. I went. Each oak was encircled by a thick wall of brush, over which it was almost impossible to see more than a fraction of the tree, and the high oak tops were impenetrable to eye and glass. After chasing phantoms all the afternoon I went home with renewed respect for Billy as an adjunct to field work. In order to locate anything in chaparral, one must be high enough to overlook the mass.

That afternoon I saw a pair of phainopeplas fly up a canyon on the east, and another pair fly up another on the west. If I were to know anything of these birds, I must not be balked by faulty observing; I must at least do intelligent work. Riding in from the back and tying Billy out of sight away from the old nest, I swung myself up into a crotch of a low oak from which I could overlook the whole island. The phainopeplas soon flew in, but to the opposite side, and I was condemning myself for having driven them away when, to my amazement, the male flew over and shot down into the little oak where he had been building before! My self-reproach took a different form — I had not been patient enough. Surely if I could wait an hour for an ordinary hummingbird, I could wait a morning for an absent phainopepla.

From the nest the beautiful bird flew to the bare oak top behind it which he used for a perch, and — alas! gave his warning call. I was discovered. He dashed his tail, turned his head to look at me first from one side and then from the other, and then flew to the top of the highest tree in sight to verify his observations. Whether he recognized the object as his pepper-tree acquaintance, I do not know; but to my great relief he went back to his work. By this time the little tree which had seemed such a comfortable chair had undergone a change — I felt as if stretched upon the gridiron of St. Anthony. Climbing down stiffly, I kneeled behind the brush and practiced focusing my glass on the nest so that it would not catch the light and frighten the bird, when out he flew from the nest and sat down facing me in broad daylight! He did not say a word, but looked around abstractedly, as if hunting for material.

If he were so indifferent, perhaps it would be safe to creep nearer. Following the paths trodden by the bare feet of the school children, and spying and skulking, I crept into a good hiding-place about a rod from the nest. The ground was covered with dead leaves, and I saw a suggestive round hole — a very large rattlesnake had been killed a few rods away the week before. I covered the hole with my cloak and then sat down on the lid — nothing could come up while I was there, at all events.

The phainopepla worked busily for some time, flying rapidly back and forth with material. Then came the warning cry. I drew in my notebook from the sun so that it should not catch his eye, and waited. The hot air grew hotter, beating down on my head. A big lizard wriggled over the leaves, and I thought of my rattlesnake. Then Billy sneezed in a forced way, as though to remind me not to go off without him. Growing restless, I moved the bushes a little — they were so stiff they made a very good chair-back if one got into the right position — when suddenly, looking up I saw my phainopepla friend vault into the air from a bush behind me, where, apparently, he had been sitting taking notes of his own! What observers birds are, to be sure! The best of us have much to learn from them.

But though the phainopepla was most watchful, he was open to conviction, and he and his mate at last concluded that I meant them no harm. Afterwards, when I moved, they both came and looked at me, but went about their business quite unmindful of me.

As I had seen from the outset, the male did almost all the building. When his spouse came in sight he burst out into a tender joyous love song. She went to the nest now and again, but generally when she came it was to sun herself on the bare perch tree, where she dressed her plumes or merely sat with crest raised and her soft gray feathers fluffed about her feet, while waiting for her mate to get leisure to take a run with her.

When he had finished his stint and she was not about, he would take his turn on the perch tree, his handsome glossy black coat shining in the sun. If an unwitting neighbor lit on his tree he would flatten his crest and dash down indignantly, but for the most part he perched quietly except to make short sallies into the air for insects, sometimes singing as he went; or he just warbled to himself contentedly, what sounded like the chattering run of a swallow on the wing. One day we had quite a conversation. His simplest call note was like the call of a young robin, and while I answered him he gave his note seventeen times in one minute, and eleven times in the next half minute.

The birds had a great variety of calls and songs, most of which were vivacious and cheering and seemed attuned to the warmth and brightness of the California sunshine. The quality of the love song was rich and flute-like.

The male phainopepla seemed to enjoy life in general and his work in particular. He frequently sang to himself when going for material; and once, apparently, when on the nest. When he was building I could see his black head move about between the leaves. Like the gnatcatchers, he used only fine bits of material, but he did not drill them in as they did. He merely laid them in, or at most wove them in gently. Now and then, as the black head moved in front, the black tail would tilt up behind at the back of the nest as if the bird were moulding; but there was comparatively little of that. When completed, the nest was a soft felty structure.

When working, the male would fly back and forth from the ground to the nest, carrying his bits of plant stem, oak blossom, and other fine stuff. He worked so rapidly that it kept me busy recording his visits. He once went to the nest four times in four minutes; at another time, seventeen times in a little over an hour. Sometimes he stayed only half a minute; when he stayed three minutes, it was so unusual that I recorded it. He worked spasmodically, however. One day he came seventeen times in one hour, but during the next half hour came only five times. The birds seemed to divide their mornings into quite regular periods. When I awoke at half past five I would hear them at the pepper-trees breakfasting; and some of them were generally there as late as eight o'clock. From eight to ten they worked with a will, though the visits usually fell off after half past nine. It was when working in this more deliberate way that the male would go to his perch on an adjoining tree and preen himself, catch flies, or sing between his visits. Once he sat on the limb in front of the nest for nearly ten minutes. By ten o'clock I found that I might as well go to watch other birds, as little would be going on with the phainopeplas; and they often flew off for a lunch of peppers.

Just as the island nest was about done — it was destroyed! I found it on the ground under the tree. For a time I felt as if no nests could come to anything; the number that had been destroyed during the season was disheartening. It seemed as though I no sooner got interested in a little family than its home was broken up. Sometimes I wondered how a bird ever had courage to start a nest.

But though it was hard to reconcile myself to the destruction of the phainopeplas' nest, I found others later. Altogether, I saw three pairs of birds building, and in each case the male was doing most of the work. Two of the nests I watched closely, watch and note-book in hand, in order to determine the exact proportion of work done by each bird. One nest was watched two hours and a half, during a period of five days, in which time the male went to the nest twenty-seven times, the female, only three. The other nest was watched seven hours and thirty-five minutes, during a period of ten days, in which time the male was at the nest fifty-seven times; the female, only eight. Taking the total for the two nests: in ten hours and five minutes the male went to the nest eighty-four times; the female, eleven. That is to say, the females made only thirteen per cent of the visits. In reality, although they went to the nest eleven times, the ratio of work might safely be reduced still further; for in watching them I was convinced that, as a rule, they came to the nest, not to build, but to inspect the building done by their mates. Indeed, at one nest, I saw nothing to make me suspect that the female did any of the work. Her coming was usually welcomed by a joyous song, but once the evidence seemed to prove that she was driven away; perhaps she was too free with her criticisms! In another case the work was sadly interrupted by the presence of the visitor, for while she sat in the nest her excited mate flew back and forth as if he had quite forgotten the business in hand. Perhaps he was nervous, and wanted to make sure what she was doing in the new house!

In several instances I found that while the males were at work building, the females went off by themselves. Once I saw Madame Phainopepla bring her friend home with her. No sooner had the visitor lit than — shocking to relate — the lord of the house left his work and drove her off with bill and claw — a polite way to treat his lady's friends, surely! On one occasion, when I looked up I saw a procession passing overhead — two females followed by a male. The male flew hesitatingly, as if troubled by his conscience, and then, deciding that if the nest was ever going to be built he had better keep at it, turned around and came back to work. One day when I rode over to the chaparral island, I found two of the males sitting around in the brush. They played tag until tired, and then perched on a branch in the sun, side by side, evidently enjoying themselves like light-hearted, care-free bachelors. Their mates were not in sight. But suddenly I glanced up and saw two females flying in to the island high overhead, as if coming from a distance. Instantly the indifferent holiday air of their mates vanished. They gave their low warning calls, for I was on the ground and they must not show me their nests. In answer to the warning the females wavered, and then, when their mates joined them, all four flew away together.

At other times when I rode in the males would make large circles, seventy-five feet above me, as if to get a clear understanding of the impending danger. This was when small nest hunters were about, and the birds were some whose nests I did not find, and who had no opportunity to become convinced of my good intentions.

After finding that the males did most of the building, I was anxious to see how it would be when the brooding began. Three of my nests were broken up beforehand, however, and the fourth was despoiled after I had watched the birds on the nest one day. Nevertheless, the evidence of that day was most interesting as far as it went. It proved that while the female lacked the architect's instinct, she was not without the maternal instinct. There were two eggs in the nest, and in the one hour that I watched, each bird brooded the eggs six times. Before this, the female had been to the nest so much less than the male that now she was much shyer; but although Billy frightened her by tramping down the brush near by, it was she who first overcame her fears and went to cover the eggs.

XIX.

MY BLUE GUM GROVE.

ONE of the first things I did on getting settled on my ranch, the second time I was in California, was to get a wagon and go down to my eucalyptus grove for a load of the pale green aromatic boughs with which to trim my attic study; for their fragrance is delightful and their delicate blue-green tone lends itself readily to decorative purposes. When the supply needed replenishing, I rode down on Mountain Billy and carried home the sweet-smelling branches on the saddle.

The grove served a more utilitarian purpose, however. The eucalyptus is an Australian tree, with narrow straight-hanging leaves, and its rapid growth makes it useful for firewood. A tree will grow forty feet in four years, and when cut off a few feet above the ground will spring up again and soon be ready to yield another crop. My grove had never been cut, but would soon be old enough. In the photograph of a eucalyptus avenue near Los Angeles, the row of trees on the right have been cut near the ground and the branching trunks are the consequence.

My eucalyptus or blue gum grove was down near the big sycamore, and opposite the bare knoll where Romulus and the burrowing owls had their nightly battles. On one side of it was a rustling cornfield always pleasant to look at. After the bare yellow stubble and all the reds and browns of a California summer landscape, its rich dark green color and its stanch strong stalks made it seem a very plain honest sort of field, and its greenness was most grateful to eyes unused to the bright colors and strong lights of California.

Opposite the little grove, in a small house perched on a hill, an old sea-captain lived alone. As I rode by one day, he sat with his feet hanging over the edge of the high piazza, looking off; as if on the prow of his vessel, gazing out to sea. When I stopped to ask if he had seen anything noteworthy happen at the grove, he complained that it shut off his view and kept away the breeze from the ocean! I was too much taken by surprise to apologize for my trees, but felt reproached; unwittingly I had destroyed the old captain's choicest pleasure. He had spoken in an impersonal way that I quite understood, — he had been taken unawares, — but the next time I rode past, as if to make up for any apparent rudeness, he came hurrying down the walk to tell me of a crow's nest he had seen in the grove. To mark it he had

EUCALYPTUS AVENUE, SHOWING POLLARDED TREES ON THE RIGHT, NEAR LOS ANGELES

fastened a piece of paper to the wire fence by the road, and another paper to the nest tree, binding it on with a eucalyptus twig in true sailor fashion.

It was always a relief to leave the hot beating sun and the glare of the yellow fields and enter the cool shade of the quiet grove. I could let down the fence and put it up behind me; thus having my small forest all to myself; and used to enjoy riding up and down the fragrant blue avenues. The eucalyptus-trees, although thirty or forty feet high, were lithe and slender; some of them could be spanned by the hands. The rows were planted ten feet apart, but the long branches interlaced, so one had to be on the alert, in riding down the lines, to bend low on the saddle or push aside the branches that obstructed the way. The limbs were so slender and flexible that a touch was enough to bend back a green gate fifteen to twenty feet long, and Billy often pushed a branch aside with his nose. In places, fallen trees barred our path, but Billy used to step carefully over them.

The eucalyptus-trees change very curiously as they grow old. When young they are covered with branches low to the ground, and their aromatic tender leaves are light bluish green; afterwards they lose their lower branches, while their leaves become stiff and sickle-shaped, dull green and almost odorless. The same changes are seen in the bark: first the trunks are smooth and green; then they are hung with shaggy shreds of bark; this in turn drops off so that the old trees are smooth again. Some of the young shoots have almost white stems, and their leaves have a pinkish tinge. Indeed, a young blue gum is as pretty a sight as one often sees; it is a tree of exquisite delicacy of coloring.

Mountain Billy and I both liked to wander among the blue gums. Billy liked it, perhaps, for association's sake, for we had ridden through the eucalyptus at his home in northern California. I too had pleasant memories of the northern gums, but my first interest was in finding out who lived in my little woods. A dog had once been seen driving a coyote wolf out of it, but that was merely in passing. I did not expect to meet wolves there. It was said, however, to be a good place for tarantulas, so at first I stepped over the dead leaf carpet with great caution; but never seeing any of the big spiders, grew brave and sat indifferently right on the ground before the nests, or leaning up against the trees. The ground was almost as hard as a rock, for the eucalyptus absorbed all the moisture, and that may have had something to do with its freedom from snakes and scorpions, though it would not explain the absence of caterpillars and spiders, which just then were so common outside. Though in the

EUCALYPTUS WOOD STORED FOR MARKET, IN A EUCALYPTUS GROVE
NEAR LOS ANGELES

grove a great deal, I never ran into but one cobweb, and was conscious of the pleasant freedom from falling caterpillars. Moreover, I never saw a lizard in the blue gums, though dozens of them were to be seen about the oaks and in the brush.

It was a surprise to find so many feathered folks living in the eucalyptus, and I took a personal interest in each one of the inhabitants. The first time we started to go up and down the avenues we scared up a pair of turtle doves, beautiful, delicately tinted gentle creatures, fit tenants of the lovely grove. They did not know my friendly interest in them, and flew to the ground trailing and trying to decoy me away in such a marked manner that when we passed a young dove a few yards farther on, it was easy to put two and two together.

Yellow-birds called *cheet'-tee*, *ca-cheet'-ta-tee*, and the grove became musical with the sweet calls of the young brood. There was one nest with a roof of shaggy bark, and I wondered if the birds thought it would be pleasant to live under a roof, or whether the bark had fallen down on them after they built. I could get no trace of the owners of the nest, and it troubled me, not liking to have any little homes in my wood that I did not know all about. As we went down one aisle, a big bird went blundering out ahead of us, probably an owl, for afterwards we stumbled on a skeleton and feathers of one of the family.

In one of the trees we came to an enormous nest made of the unusual materials that are sometimes chosen by that strange bird, the road-runner. It was an exciting discovery, for that was before the road-runner had come to the ranch-house, and I had been pursuing phantom runners over the hills in the vain attempt to learn something about them; while here, it seemed, one had been living under my very vine and fig-tree! To make sure about the nest, I spoke to my neighbor ranchman, and he told me that when he had been milking during the spring he had often seen the birds come out of the blue gums, and had also seen them perching there on the trees. How exasperating! If I had only come earlier! Now they had gone, and my chance of a nest study was lost.

But my doll was not stuffed with sawdust, for all of that. There was still much to enjoy, for a mourning dove flew from her nest of twigs almost over Billy's head, and it made me quite happy to know that the gentle bird was brooding her eggs in my woods. Then it was delightful to see a lazuli bunting on her nest down another aisle. It seemed odd, for there was her little cousin nesting out in the weeds in the bright sun, while she was raising her brood in the shady forest. The two nests were as unlike as the sites. The bird outside had used dull green weeds, while this one used beautiful shining oak stems. I thought the pretty bird would surely be safe here, but one day when I called, expecting to see a growing family, I was shocked to find a pathetic little skeleton in the nest.

One afternoon in riding down the rows, I came face to face with two mites of hummingbirds seated on a branch. Their grayish green suits toned in with the color of the blue gums. It was a surprise when one of them turned to the other and fed it — the mother hummer was small enough to be taken for a nestling! She sat beside her son and fed him in the conventional way, by plunging her bill down his open mouth. When she had flown off, he stretched his wings, whirred them as if for practice, and then moved his bill as if still tasting the dainty he had had for supper. He sat very unconcernedly on a low branch right out in the middle of the road, but Billy did not run over him.

I found two hummers' nests in the eucalyptus during the summer. One builder was the one the photographer was fortunate enough to catch brooding; her nest, the one so charmingly placed on a light blue branch between two straight spreading leaves, like the knot between two bows of stiff ribbon.

The second nest was on a drooping branch, and, to make it stand level, was deepened on the down side of the limb, making it the highest humming-bird's nest I had ever seen. It was attached to a red leaf — to mark the spot, perhaps — one often wonders how a bird can come back twice to the same leaf in a forest. How one little home does make a place habitable! From a bare silent woods it becomes a dwelling-place. Everything seemed to centre around this little nest, then the only one in the grove; the tiny pinch of down became the most important thing in the woods. It was the castle which the trees surrounded.

When I first found the nest it held two white warm eggs about as large as peas, and I became much interested in watching their progress, often riding down to see how they were getting on. The hummer did not return my interest. She was nervous, darting off when Billy shook himself or when the shadow of a soaring turkey buzzard fell over the nest; but in spite of that we made ourselves quite at home before her door. I would dismount and sit on the ground, leaning against a blue gum, while Billy stood by, in a bower of green leaves, with ears pricked forward thoughtfully, and a dreamy look of satisfaction in his eyes. Hummingbirds are such dainty things. Once when this one alighted on the rim of her nest she whirred herself right down inside. Soon she began to act so strangely for a brooding bird that, when she flew, I went to feel in the nest. The tips of my fingers touched what felt like round balls, but, not satisfied, I pulled down the bough and found one round ball and one mite of a gray back with microscopic yellow hairs on each side of the spine. The whole tiny body seemed to throb with its heart beats. I wondered how such a midget could ever be fed, but found, as in the case of the hummer under the little lover's tree, that the mother gave its food most gently, reserving her violent pumping for a more suitable age; though one would as soon think of poking a needle down a baby's throat as that bill.

Often, while watching the nest, my thoughts wandered away to the grove itself. The brown earth between the rows was barred by alternate lines of sunlight and shadow, and the vista of each avenue ended in blue sky. Sometimes cool ocean breezes would penetrate the forest. The rows of trees, with their gently swaying, interlacing branches, cast moving shadows over the sun-touched leafy floor, giving a white light to the grove; for the undersides of the young eucalyptus leaves are like snow. From the stiff, sickle-shaped upper leaves the sun glanced, dazzling the eyes. Mourning doves cooed, and the sweet notes of yellow-birds filled the sunny grove with suggestions of happiness. A yellow butterfly wandered down the blue aisles. Such a secure retreat! I returned to it again and again, coming in out of the hot yellow world and closing behind me the doors of my 'rest-house,' for the little wood had come to seem like a cool wayside chapel, a place of peace.

And when I finally left California, deserting Mountain Billy to return to the East, of all my haunts the one left the most unwillingly was the little blue gum grove, the peaceful wayside rest-house, in whose whitened shade we had spent so many quiet hours together.

INDEX.

INDEX TO ILLUSTRATIONS.